Unprecedented Evolution

Unprecedented Evolution

Continuities and Discontinuities between Human and Animal Life and the Future of Humanity

Spyridon A. Koutroufinis
& René Pikarski
Editors

ANOKA, MINNESOTA 2020

Unprecedented Evolution: Continuities and Discontinuities between Human and Animal Life and the Future of Humanity

© 2020 Process Century Press

All rights reserved. Except for brief quotations in critical publications and reviews, no part of this book may be reproduced in any manner without prior permission from the publisher.

Process Century Press
RiverHouse LLC
802 River Lane
Anoka, MN 55303

Process Century Press books are published in association with the International Process Network.

Cover design: Susanna Mennicke

Image Credit, Pan and Daphnis: http://www.flickriver.com/photos/mharrsch/1690023097/

Volume XXI
Toward Ecological Civilization Series
Jeanyne B. Slettom, Series Editor

ISBN 978-1-940447-47-6
Printed in the United States of America

SERIES PREFACE: TOWARD ECOLOGICAL CIVILIZATION

We live in the ending of an age. But the ending of the modern period differs from the ending of previous periods, such as the classical or the medieval. The amazing achievements of modernity make it possible, even likely, that its end will also be the end of civilization, of many species, or even of the human species. At the same time, we are living in an age of new beginnings that give promise of an ecological civilization. Its emergence is marked by a growing sense of urgency and deepening awareness that the changes must go to the roots of what has led to the current threat of catastrophe.

In June 2015, the 10th Whitehead International Conference was held in Claremont, CA. Called "Seizing an Alternative: Toward an Ecological Civilization," it claimed an organic, relational, integrated, nondual, and processive conceptuality is needed, and that Alfred North Whitehead provides this in a remarkably comprehensive and rigorous way. We proposed that he could be "the philosopher of ecological civilization." With the help of those who have come to an ecological vision in other ways, the conference explored this Whiteheadian alternative, showing how it can provide the shared vision so urgently needed.

The judgment underlying this effort is that contemporary research and scholarship is still enthralled by the seventeenth century view of nature articulated by Descartes and reinforced by Kant. Without freeing our minds of this objectifying and reductive understanding of the world, we are not likely to direct our actions wisely in response to the crisis to which this tradition has led us. Given the ambitious goal of replacing now dominant patterns of thought with one that would redirect us toward ecological civilization, clearly more is needed than a single conference. Fortunately, a larger platform is developing that includes the conference and looks beyond it. It is named Pando Populus (pandopopulous.com) in honor of the world's largest and oldest organism, an aspen grove.

As a continuation of the conference, and in support of the larger initiative of Pando Populus, we are publishing this series, appropriately named "Toward Ecological Civilization."

~John B. Cobb, Jr.

OTHER BOOKS IN THIS SERIES
An Axiological Process Ethics, Rem B. Edwards
Panentheism and Scientific Naturalism, David Ray Griffin
Organic Marxism, Philip Clayton & Justin Heinzekehr
Theological Reminiscences, John B. Cobb, Jr.
Integrative Process, Margaret Stout & Jeannine M. Love
Replanting Ourselves in Beauty, Jay McDaniel & Patricia Adams Farmer, eds.
For Our Common Home, John B. Cobb, Jr., & Ignacio Castuera, eds.
Whitehead Word Book, John B. Cobb, Jr.
The Vindication of Radical Empiricism, Michel Weber
Intuition in Mathematics and Physics, Ronny Desmet, ed.
Reforming Higher Education in an Era of Ecological Crisis and Growing Digital Insecurity, Chet Bowers
Protecting Our Common, Sacred Home, David Ray Griffin
Educating for an Ecological Civilization, Marcus Ford & Stephen Rowe, eds.
Socialism in Process, Justin Heinzekehr & Philip Clayton, eds.
Two Americas, Stephen C. Rowe
Rebuilding after Collapse, John Culp, ed.
Putting Philosophy to Work, John B. Cobb Jr. & Wm. Andrew Schwartz, eds.
What Is Ecological Civilization, Philip Clayton & Andrew Schwartz
Conceiving an Alternative, Demian Wheeler & David Conner, eds.
Rethinking Consciousness, John H. Buchanan & Christopher Aanstoos, eds.

Dedication

For John B. Cobb, Jr. and Philip Clayton, with admiration and gratitude for their inspired work and grave concern for the fate and future of our planet.

About the cover: Why Pan and Daphnis?

The sculpture on the cover shows the ancient Greek god Pan (Πάν) teaching his beloved friend, the young shepherd Daphnis, to play the pan flute. The sculpture is a Roman copy of the Greek original, circa 100 BCE, that was found in Pompeii. In ancient Greek religion and mythology, Pan is worshiped as the god of the wild—nature, mountains, fields, groves, wooded glens, but also shepherds and flocks and rustic music. Because of his affiliation with vigorous and impulsive sexuality, he is connected to fertility and the season of spring.

Pan was usually worshipped in caves and grottoes. The origin of his worship traces back deep in mythic time. Like other deities of nature, Pan appears to be older than the Olympian gods of ancient Greek religion. In orphic and stoic pantheism, Pan was raised to a primordial god, the god of everything. This reflects the folk etymology that equates Pan's name with the Greek word for "all" ("pan," πᾶν). With the hindquarters, legs, and horns of a goat, his body combines harmonically human and animal parts. We, the editors, recognize in his stature an allegory of the rootedness of humankind in the animal kingdom, and thus consider Pan a symbol of the continuity between human and animal life, to which we have dedicated a significant part of the present volume.

We think that, despite the human parts of Pan's stature, Daphnis' well-formed human body contrasts to the animality of the primordial god. Daphnis' coy posture builds a clear opposition to Pan's impulsiveness, and rampant desire. This obvious polarity represents the undeniable discontinuity of cultivated humanness to the vigorous instincts of wild animal nature, which is another important subject of this book. Nevertheless, despite his essential distance to Pan's elemental nature, the shepherd Daphnis, himself a child of nature, is receptive for his immortal friend's sympathy and his gift—the teaching to play the flute.

We recognize in the idealistic profession of sympathy of the primordial god of nature for his young and inexperienced human friend another original allegory for the continuity between human and all nonhuman life. We consider the friendship of Pan and Daphnis a symbol of the genuine human ability to establish anew an intuitive, non-analytic contact with the unspoiled wisdom of nature, which is represented in our cover by the Earth globe in the background. This contact will gift humankind with the emergence of a higher and unprecedented form of our lost primordial innocence, which is ideally symbolized by the play of Pan's flute. The present volume owes its origin to that intuition.

TABLE OF CONTENTS

INTRODUCTION, 1
Anthropogenesis as an Evolution of Contrasts
in Need of Harmonization
Spyridon A. Koutroufinis and René Pikarski

ONE, 11
Humanness
Terrence W. Deacon

TWO, 19
Religion in Human and Cosmic Evolution:
Whitehead's Alternative Vision
Matthew T. Segall

THREE, 35
"Are We Human or Are We Dancer?"
René Pikarski

FOUR, 63
Evolution and what the Intellect Makes of it
Alex Gomez-Marin

FIVE, 91
Human vs. Animal Relation to the Environment: A
Bergsonian-Whiteheadian Perspective on Uexküll's
Concept of 'Umwelt' and Cassirer's 'Animal
Symbolicum'
Spyridon A. Koutroufinis

SIX, 123
Views of Future Human Evolution and the Future Human:
Technology and Transhumanist-Based Perspectives that
Separate Humanity from the Animal Kingdom
Linda Groff

SEVEN, 155
Regarding Humanism: Some Observations Concerning the
Tibetan Buddhist and Transhumanist Dialogue
Sean K. MacCracken

EIGHT, 185
Shared Vulnerabilities: Cosmic Consciousness and the
Philosophy of Organism
Jason James Kelly

NINE, 205
Technosophia: A Cosmohumanist Manifesto
Theo Badashi

ENDNOTES, 243
CONTRIBUTORS, 285
Index of Persons, 287
Index of Concepts, 291

INTRODUCTION

Anthropogenesis as an Evolution of Contrasts in Need of Harmonization[1]

Spyridon A. Koutroufinis and René Pikarski

LIFE TAKES MANY FORMS. Human beings can be viewed as just one of these lifeforms among others. Although it is certainly true that humans are one biological species among many others, it is also the case that humans are different from all the others in quite remarkable ways. We have produced works of exquisite beauty, sophistication, and enduring value. And yet at the same time we are threatening to drive ourselves as well as untold numbers of other species on Earth into extinction. To make sense of this anthropogenic paradox we first need to understand how we diverged from the rest of our animal cousins in the course of our evolution. We need to understand this evolutionary development, both in terms of how we are continuous with other species and how we differ. There is not only diversity in the *nature* of animal and human evolution, but likewise in the *stories* we tell about it: a diversity of scientific explanations and speculative understandings of continuities and discontinuities between animal and human life.

The aim of this book is to provide a dynamic hub where the different angles, modes, and contents of these stories can contact one another and be discussed. The chapters emerged out of the anthropological section that we, the editors, organized for the international conference "Seizing an Alternative: Toward an Ecological Civilization" that took place in Claremont, CA (June 2015). The authors included in this volume examine core dimensions of *conditio humana* in light of *process philosophy*. They apply process metaphysics to core anthropological issues, including the survival of both the human species and the biosphere as a whole. With a general focus on the unique capacity for symbolization as marking an important and influential factor in human evolution, the following specific questions frame the different chapters of this book:

- How did symbolic thought shape the evolution of the human species?
- How did symbolic systems shape human experience of and reasoning about space, time, matter, life, and natural processes?
- How do our unique forms of power relations distinguish humans from other species?
- How do our spiritual and metaphysical belief systems influence human rationality and morality?
- How can we balance our spiritual needs with our rational abilities, and how could this influence our future evolution?
- How should we respond to the trends towards transhumanism and bio-technocracy?

Without neglecting the evolutionary continuity between the human species and the animal kingdom, the authors reflect on human particularities compared to animals in respect to the past as well as the future of our species. Considering the clear and present danger caused by human actions on Earth, all of the authors agree that retrospective as well as prospective thoughts and intuitions regarding human particularities must be guided by an overriding sense of responsibility, not only for our actions, but to understand our past and shape the future of human evolution. Two important facts about human nature moved us to edit this book. The first one is that anthropogenesis gave rise to a highly antagonistic contrast between a *need* and a *faculty*, both of which are uniquely characteristic of human consciousness. This contrast amounts to the most essential difference between humans and animals.

Humans need to make *sense* of their activities. The more our activities find their place within a social environment, the more they mean something to us. We live by values which we, whether consciously or unconsciously, permanently reinforce or undermine through our activities, thoughts, emotions, and desires. This is true especially of ethical values. The evolutionary transition from animal to human being is inseparably connected to the emergence of a new form of creativity: *freedom*. Even though freedom is often attributed to animals, animal creativity does not reach the level of freedom. The latter should not be reduced to the ability to choose between different alternatives of action that, of course, must be assigned even to animals with simple nervous systems. Freedom can only be ascribed to beings that feel the need to justify their actions within the framework of ethical values and that have the ability to do so. Freedom is creativity that is indissolubly bound to sociality, that is to say, creativity that is guided by principles, directed towards the creation of meaning and sense, and guided by values and ethical considerations. Therefore, instead of attributing freedom to animal actions, it would be more appropriate to talk of animal creativity.

Besides the need to realize ethical sociality, our species stands out due to our highly developed faculty to think abstractly. Although many animal species manifest the ability of abstract thinking, in human beings the evolution of conceptual reasoning has reached an unprecedented peak. This has enabled us to build universally usable tools. The most powerful of them are our abstract concepts and the symbolic systems (e.g., scientific theories and social utopias) into which they are integrated. Both allow us not only to build increasingly complex material tools but also to invent narrative structures, such as myths in a broad sense of that term. Myths, for example, reaffirmations of "faith" in endless scientific and technological progress as well as economic growth, rationalize the desire to manipulate nature by employing material tools and, even more importantly, justify our right to do so. One of the most powerful instruments that emerged in the evolution of the human mind is modal thought: the ability to combine abstract concepts in order to imagine highly complex potential future situations, which in contemporary philosophy are referred to as "possible worlds."

Due to operating with abstract concepts that can be combined in unlimited ways one with the other, human thinking is able to create an endless number of propositions that outline the image of future possible actions. This mental faculty corresponds with the evolution of both episodic (experiential) and semantic (propositional) memory. At some time in the Pleistocene, however, due to their capacity to anticipate the future and remember the past, our distant ancestors became aware of the limitedness of their existence. If beings with highly developed mental faculties and a strong feeling of individuality become aware of their mortality, it is only a matter of time before different forms of desire for immortality and a thirst for eternity arise. Our faculty of conceptual thinking opened to us ways of canalizing different forms of mental and physical work that enabled us to create highly complex works that have outlived their creators by centuries and millennia. Philosophical, scientific, political, and economical works, such as the Vedas, the works of Plato, Hegel, Newton, Adam Smith, and Karl Marx, as well as material creations, such as the pyramids, ancient Greek temples and sculptures, and the Moai statues of the Easter Island, witness a conscious or unconscious desire of their creators to reach eternity. There is no question that most of these enormous achievements continue to perform a great service for humankind. Highly complex abstract or material creations harbor, however, a significant risk: they can become ends in themselves. This applies especially to our theoretical systems, because abstract structures tend to seduce their creators and their users into removing them from the social and ethical contexts within which they were developed. As a result, political visions; economic growth (money is also an abstract medium); work; highly abstract systems of knowledge; technological achievements; faith; the exercise of different forms of power; and fine arts have the virtually irresistible tendency to become ends in themselves. However, activities that have become self-referential ends in themselves increasingly lose contact with reality, i.e., to the wider socio-ethical context. Sooner or later this detachment from reality ends in disasters of different dimensions because self-purposes cannot limit their own expansion before they cause the collapse of their carrier and thus of themselves. In the last thirty years we have witnessed two events of historical proportions that can be traced back to stubborn insistence of political and financial elites on their beloved convictions: the collapse of real existing socialism and the enormous financial crisis of 2008 that is the worst economic disaster since the Great Depression of 1929.

We think that there is a severe antagonism between the need for meaningful social embeddedness, on the one hand, and on the other, the strong tendency of our intellectual ability to give birth to different sorts of abstract and material systems that become ends in themselves, threatening social togetherness. This antagonism is the most fundamental struggle that impels human evolution. The increasing frequency and severity of all kinds of crises have made obvious that this antithesis must become balanced if the human species is to survive. Thus, the second central thought motivating this book is the urgent necessity of raising the aforementioned antagonism characterizing human nature to a new quality of harmonization before it is too late.

Over the last seventy years humankind has presided over the most severe ecological crisis in 65 million years. It is obvious that this crisis is rooted in the central myth of Western civilization that consists of three interconnected dimensions: the belief in endless economic growth, technological manipulation of the environment, and scientific reduction of living and inorganic nature. (The latter reductionistic perspective is the basis upon which biotechnological manipulations of vast agricultural areas are undertaken, manipulations that will likely have severe ecological consequences in the future). This tripartite myth is the most solidified end in itself that dominates the destiny of the entire human species at the present. What nevertheless makes our century so outstanding is that we must establish a new balance not only within the framework of our globalized human world but rather inside the whole living environment. We must expand our sense of ethical sociality throughout the whole living environment: Earth's biosphere. It is self-evident that this must occur not only in a spatial but also in a temporal sense. We must learn to justify our actions within an ethical framework that is seriously concerned about the present and *future* biosphere of Earth.

Following this imperative a central topic of this book is the future of human evolution. In which directions could human activities canalize our physical and mental evolution? Will we be able to integrate our intellectual abstractions with our intuitive insights so as to bring forth a new synthetic consciousness that will raise humanity to a beneficial evolutionary factor in the biosphere? Several chapters in this volume offer new approaches to ecology, science, spirituality, and technology as a counter project to the vision of the future provided by *scientistic-technocratic transhumanism*. Although most contemporary transhumanistic visions, such as the achievement of personal immortality or the creation of a new human species adapted to space travel and life on other planets, are being discussed only within limited circles, they must be taken seriously by all who are concerned about the future of Earth and our species. For just as our visions and hopes provide information about our unconscious, transhumanistic visions reveal the implicit beliefs of experts, many of whom work for big biotechnological and other high-tech corporations, about the human body and corporeality in general. Transhumanistic visions are much more than they seem to be. They display the unconscious fundamental attitudes of proponents of scientistic technocracy in its totality. Thus, even if those visions sound bizarre, we think that they must be considered as a grave "radiograph" of the mind and soul of a powerful elite that extends far beyond the transhumanist community and might dominate a near-future society by enforcing scientism and technocracy. Hence, rather than being dismissed as the grotesque thoughts of "nerds," transhumanistic visions should be taken seriously as warning signals about just how alienated our technologically shaped species has become from the fragile natural world we depend upon.

To show how process philosophies inspired by Henri Bergson and Alfred North Whitehead can provide a diversity of fertile responses to these issues, we have collected the views of nine thinkers from different countries and discourses.

For us, there is no better way to start this volume than with a chapter called "Humanness." What if our biological nature cannot be separated from our symbolic nature? And if we are the "symbolic species," then how does the human capacity for symbol use provide us with a clue to understanding our own evolution? *Terrence W. Deacon* presents his idea of an inextricable connectivity between two modes of social transmission in human evolution. He argues that a "non-genetic-non-physiological" interaction of semiotic processes is largely responsible for the way our organism, and especially our brains, evolved. A symbolic component of social inheritance is interwoven with a nonsymbolic social-cognitive inheritance. Deacon talks about three crucial points of his research: How does the importance of symbols for our evolution challenge us to find a new and more adequate theory of symbolic reference? What does it mean that humanness is primarily based on an unprecedented semiotic dynamic that influences our nervous system and its functions? After all, which mental predispositions make us specifically human?

Proceeding from Deacon's discussion of the extraordinary importance of symbols for marking human continuities and discontinuities with animal life, the chapter "Religion in Human and Cosmic Evolution" focuses on those processes where symbolization may find its most powerful manifestation: in humanity's diverse expressions of spirituality, mythmaking, and religion. *Matthew T. Segall* investigates the evolutionary origins of human religion according to the thoughts of Robert Bellah and Alfred North Whitehead. What does it mean that, far beyond any static definition, human religion and spiritualty are interrelated dimensions of an ongoing, cosmologically-emergent activity? How does this cosmological view of the relationship between scientific theorization and religious mythopoeia reveal the important role of our capacity for spontaneous playfulness? What was the role of play in humanity's evolutionary past, and how might a recovery of this playfulness influence the future of human evolution? Segall argues that religion and spirituality are far more than just accidental products of blind biological forces; they are expressions of the unfolding creativity of the universe in its human mode.

Indeed, a better understanding of the human relation to the universe requires a continual integration of biological and social perspectives. In this process, the human condition itself does not remain static. *René Pikarski* reflects on how the dynamic interplay of biology and sociology depends upon our view of life and power. On this account, "Are We Human or Are We Dancer?" attempts to bring together the thoughts of Henri Bergson and Michel Foucault. What does it mean that our intelligent and intuitive performances tend towards a high degree of flexibility, openness, and universality within the configuration of our relations with and references towards the world, our environment, others, and ourselves? And what does it mean that these capabilities and performances intrinsic to our condition usually take place in social frameworks and therefore in the context of power relations? Pikarski argues that a dynamic understanding of the human condition refers, at its core, to the human process of becoming a subject by developing a self that critically reflects on and transforms the dependencies that surround and penetrate it. Thus, the "care of the self" is a further expression of human creativity that, unfortunately, is often inhibited by the power mechanics of our own life-denying form of intellect.

Alex Gomez-Marin begins his chapter, "Evolution and What the Intellect Makes of It," with an elegant provocation: The human being is caught in a paradox, since we think of ourselves not only as a *result* of evolution, but also as the *developer* of various theories of evolution. In other words, the evolutionary process, in an unprecedented attempt, has been thought by one of its products: the explanandum has nominated itself as the explanans. Nevertheless, symbolizing, conceptualizing, thinking, and speaking about evolution is one thing, while evolution itself is another. Given this gap, we should be worried about the way our intellectual discourses and capacities have lost contact with the real and essential virtue of evolutionary processes. How may the intuitive thoughts of Henri Bergson on heterogeneous continuity and his notion of multiplicity be able to help us recover this essence of evolution which is veiled by our habitual ways of thinking about it?

The search for an adequate understanding of evolutionary processes further leads to the problem of how we should interpret the interactions between living beings and their environment. In his chapter, "Human vs. Animal Relation to the Environment," *Spyridon A. Koutroufinis* introduces Jakob von Uexküll's concept of *Umwelt* (environment) as a notion that subsumes several aspects of an animal's surroundings that are meaningful to it. Uexküll created a biology of subjects. With respect to human subjectivity, Ernst Cassirer expanded Uexküll's *Umwelt* by adding the component of symbolic forms and defining the human as an *animal symbolicum*. According to Cassirer, whereas animals use signs, humans communicate with symbols, which are signs that have objective meaning. It is this distinction that characterizes the specific difference between human and animal intelligence. Humans, compared to animals, not only experience the spatiotemporal aspects of their *Umwelt* in a radically different way—they live in a multidimensional symbolic *Umwelt*. Obviously, the abstract symbolic systems of contemporary physics, life science, technology, and economy enable us to act on other living beings and on the whole planet with an emotional distance, which can be destructive. Koutroufinis claims that ecologic disaster forces us to reinterpret the term *Umwelt* in order to enhance it with an ethical dimension. Such a move, he argues, requires a view of nature based on Whitehead's understanding of natural beings as embodiments of intrinsic values. Building on Bergson's concept of intuition and his advocacy for the rising of the age of intuition, Koutroufinis introduces the concept of *sacred environment*.

Now we are at a point where intuition more than intellect has guided us to the past and present aspects of human continuities and discontinuities with animal life. Thus, the chapter on *Umwelt* opens our view and discussion to the question of how we shape our environment and how it is going to shape us in the future. *Linda Groff* outlines various "Views of Future Human Evolution and the Future Human." Different technological and transhumanist perspectives that separate humanity from the animal kingdom often express the heavy impact that cultural and technological factors can have on biological evolution. Furthermore, the values of these perspectives should be critically investigated in order to determine the extent to which they are prepared to meet the challenge of creating a sustainable and living society on planet Earth—or, perhaps, on Mars. What infotech or biotech perspectives and models of transformation are on the table and with the help of which normative critera are we able to counterbalance dangerous tendencies towards a future anthropogenesis? We desperately need an evolution of human consciousness to a dynamic, interdependent, complex, and whole-systems thinking.

Expanding on this point, the next chapter, written by *Sean MacCracken*, is concerned with the future of humanity and, to this end, critically investigates transhumanism. "Regarding Humanism: Some Observations Concerning the Tibetan Buddhist and Transhumanist Dialogue" takes the nascent dialogue between both as the occasion to reexamine normative interpretations of Derek Parfit made by Max More—philosophical assumptions that have guided the development of transhumanism in its formative stages. It seeks to demonstrate that More's outlook represents a specific interpretation of Parfit, while Parfit himself shares a substantial philosophical vocabulary in common with Buddhist epistemology. Based on a recent body of literature examining the relationship between Parfit and Buddhism, how is it possible to furnish a response to the lack of gender and ethnic diversity evident amongst many transhumanist enthusiasts? Here, the guiding assumption is that an unyielding functionalist-materialist stance may be one of the key factors in "hard transhumanism," alienating large sections of the population. As an alternative, it is suggested that Parfit's category of the "empty question" is not incompatible with a Buddhist interpretation of personhood, via *Nāgārjuna,* wherein persons are held to be processes in unproblematic states of indeterminacy that are variously interpreted, not finally determined.

Many chapters in this volume assume that the contemporary ecological crisis on Earth possibly refers to what our next author terms a "disproportionation between technological success and ethical failure." From this angle, *Jason James Kelly* examines the ecological significance of cosmic consciousness in relation to Alfred North Whitehead's philosophy of organism. "Shared Vulnerabilities: Cosmic Consciousness and the Philosophy of Organism" deals with the question of how the concepts of cosmic consciousness and organism privilege a nondualistic or holistic idea of the subject-object relationship that affirms an ecological relationship with nature. By drawing on the work of the English author and activist Edward Carpenter, and the Canadian psychiatrist Richard Maurice Bucke, it can be shown that the ethical significance of cosmic consciousness coheres with the ecological teachings of spiritual ecology. Furthermore, the ethical and ecological significance of cosmic consciousness can be enhanced by cultivating a deeper engagement with Whitehead's philosophy of organism. Finally, Kelly points out an intuition that is at the basis of this entire volume: Whitehead's philosophy provides a metaphysical foundation to ground the ethical teachings of cosmic consciousness.

The last contribution of our volume is dedicated to what such a cosmic consciousness must be aware of: Humanity is losing its enlightened self-certainty about the future existence of humans due to our powerful artifacts and recent ecological and political crises. Thinking about a new human mode of being is therefore one of our most urgent tasks. "Technosophia: a Cosmohumanist Manifesto" explores ideas for a new way of being in the world relying on a specific cosmo-phenomenological worldview. *Theo Badashi* explores the fundamental principles and patterns of this cosmohumanism: What role can technology play in this new techno-ecological view, if we see it as both subjective ecological *and* objective material? How does a holistic cosmohumanism differ from the technocratic, naturalistic, and anthropocentric visions of transhumanism? Criticizing transhumanism need not mean we underestimate its role as one of the most important narratives about future humans. Instead, we should integrate transhumanist ideas into what Badashi calls the core components of a cosmohumanist paradigm: a universe-oriented cosmology in the form of a living universe story; the validation of all beings and modes of being; the emergence of participatory teleology; and a cosmological view of technology, called technosophia.

We assume that all the ideas, concepts and theses in this volume are a valuable contribution to questions regarding the processes of anthropogenesis. For us, it is an evolution of diversity and contrasts that must be harmonized. This harmonious balance is not achieved by simply unification, but by complex integration! This volume assembles a collection of reflections and intuitions on the specific ways we differ from other species, like our capacity to symbolize and create a helpful or dangerous distance from life, and our playfulness and proclivity for mythmaking. Our intellect is troubled by a consciousness of its own evolution, by the ecological and ethical challenges of a severely degraded environment, and it wields great symbolic and mythic fictions with the power to influence the bio-social evolution of future humans. Despite all our unique qualities, we cannot set ourselves apart from the natural heritage we share with every living being on planet Earth. Somewhere between transhumanism and cosmohumanism, we must find an ethical guide, an organismic and cosmic consciousness, and a speculative framework to manage our knowledge and our spontaneous actions towards the future.

ONE

Humanness

Terrence W. Deacon

IN MANY RESPECTS we humans have dual psychic ancestry. One lineage is continuous with our African ape cousins, tracing back from there to all primates, mammals, terrestrial vertebrates, and so on. The other lineage is not traceable through molecular genetics, but through continuous social transmission. It is our symbolic genealogy.[1] This dual inheritance perspective is not new. And in many respects it remains quite controversial to describe human mental nature in terms of parallel evolutionary processes. But the relationship I am describing should not be confused with dual inheritance theories that dichotomize genetic and social transmission processes, as this is defined in so-called biosocial evolution theories. In many respects my point is the opposite. It is the inextricable entanglement of two modes of social transmission that I wish to emphasize. It is the entanglement of the symbolic component of social inheritance with our nonsymbolic social-cognitive inheritance. This nongenetic nonphysiological interaction of semiotic processes has nevertheless influenced the evolution of nearly every aspect of our biology, including most notably our brains.

Our primate behavioral and symbolic psychic lineages had unrelated origins. They originated in epochs separated by tens of millions of years. The many overlapping cognitive, sensory-motor, and social-emotional predispositions that monkeys and apes inherited from the common ancestral anthropoid primate arose within the last 60 million years. In contrast, the unbroken lineage of symbolic information almost certainly doesn't extend back more than about 2.5 million years, and may have a far more recent origin. There is a radical incommensurability between these yoked semiotic genealogies. The ways that this has played out in the gradual emergence of humanness is what makes human nature so difficult to characterize in typical biological terms. It also dooms any effort to partition humanness into inherited and acquired (e. g., nature versus nurture) components. It is an unprecedented entanglement of genealogies that comprises a highly distinctive universally shared nature that sets us apart from all other species on Earth.

In my 1997 book I described humans as a "symbolic species," analogous to the way we might characterize birds as aerial species and dolphins as aquatic species.[2]

But unlike these ecologically defined adaptive types, the symbolic "ecology" that humans evolved to fit is not imposed from the physical environment. It is an ecosystem almost totally created by the species that occupies it. Moreover, in many respects we have adapted to a "virtual ecology." Though we eat, mate, and die in the same world as other species, we spend much of our lives in a world of our own making, both individually and collectively constructed, that no other species can enter. This has produced a "strange loop" of causality that led our ancestors down an evolutionary path that was unprecedented in the history of the Earth. It has made us into alien invaders, as though colonizing the Earth from a distant planet. And, like an invasive species introduced into a new habitat, the symbolic species has wreaked havoc with respect to the previous evolved order of things on Earth. So to fully understand how such a strange turn of evolutionary events came about and thus gain a better perspective from which to understand humanness with all its idiosyncrasies, we must make sense of the bio-semiotic anomaly that we are.

THE SYMBOL PROBLEM

A fundamental source of misunderstanding that has clouded judgment of our special nature is an unquestioned reliance on a simplified linguistic metaphor to characterize both our mental and our social worlds. It is based on one of the most common characterizations of word meaning: i.e., that it is based on arbitrary conventionally established correspondence. The father of twentieth-century linguistics, Ferdinand de Saussure, described this as a pairing of a signifier with what it signifies—a word with its meaning. This implicitly invokes a "code" metaphor, with a one-to-one mapping between two domains. In addition, the view that the complex combinatorial organization of words into sentences, claims, and narratives is likewise merely the result of mutational fiat providing innate grammar, or else the expression of mere local social convention, also adds to a deep skepticism concerning the groundedness of linguistic knowledge. The larger implication for understanding the nature of humanness is that there is no necessary reason to assume that there is any mental commonality that unites us. This is a serious barrier to understanding. It affects both psychological and social theories and assumes a discontinuity between social and biological causation.

Arbitrariness is a negative way of defining symbols. It basically tells us that neither formal likeness nor factual correlations are used as the basis for symbolic reference. But this is inadequate. It fails to specify exactly how the symbolic referential relation is established. So, even though this is a useful shorthand way of characterizing symbolic reference, it merely passes the buck, so to speak, to some assumed and undescribed means by which reference is established. In fact, all semiotic relationships include some degree of arbitrarity, because those attributes that are taken as the ground for the sign-object linkage can be chosen from many dimensions, and only some will be utilized (if any). What matters, then, is the convention-establishing process that generates the system of relations and the interpretive process that makes use of sign vehicle attributes to convey and project referential content. But when we attribute the establishment of symbolic reference to social convention we implicitly assume that symbolic reference is based on, and derives from, nonsymbolic semiotic processes. Indeed, the very meaning of convention implies convergence to a common habit due to some semiotically mediated interaction.

As the father of semiotic theory—Charles Peirce—recognized, symbolic reference involves a conventional type of sign vehicle that additionally represents its object of reference in a conventionally-mediated way.[3] Something can be considered symbolic, then, only if the property determining its relationship to what it refers to is also established by convention. Symbols are in this sense doubly conventional. So arbitrarity, by itself, is not diagnostic of symbolic reference. Nor can it be a critical defining feature of language.

Elsewhere I have argued that although innately fixed tendencies to communicate iconically or indexically (as in innate human "calls" like laughter and sobbing) can evolve by natural selection, symbolic reference cannot.[4] This is because of the displacement of sign attributes from features shared with what they represent. In other words, symbols lack the reliably repeated associations between properties that natural selection requires. This explains why there are no innate words, only innate calls. It is also the critical difference that distinguishes the separate but entangled semiotic lineages that are the evolutionary parents of humanness.

But even though there are no innate symbols, evolution has produced innate human predispositions that make the acquisition and use of symbolic communication comparatively easy. And, in turn, the regular use of symbolic communication and reasoning over our protracted evolutionary past has radically changed the ways we use even those cognitive abilities that long predated our symbolic awakening. These evolved modifications of human mnemonic, attentional, emotional, and social predispositions derive from, and contribute to, our unprecedented virtual niche—culture. This atypically modified neurology and radically restructured social-ecological context together are responsible for the unprecedented deviation of human biology from the biology of effectively every other species on Earth. As a result our brains and bodies have come to expect input from this ubiquitous semiotic ecosystem in order to develop normally. So-called feral children raised in isolation from language-using caretakers, are in this sense deprived of a kind of virtual genetics that is necessary for the development of humanness.

The point of the remainder of this essay is to explore some of the more distinctive consequences of this curious human evolutionary heritage. The challenge is to characterize the distinctive nature of humanness by taking onto account the impossibility of untangling our biological from our symbolic nature.

UNPRECEDENTED BRAINS

In my 1997 book *The Symbolic Species* I review evidence from comparative neuroanatomy suggesting that major quantitative changes in the proportions of different human brain structures compared with our ape cousins reflect adaptations for the unusual cognitive demands imposed by symbolic communication. This was not, however, only a case of brains adapting to language. Human brains and language co-evolved. Languages are continually being modified by social-cognitive selection processes favoring learnability and ease of use (via comparatively rapid historical change), and brain functions have been modified by eons of evolution to meet the special learning and production demands of language (via comparatively slow evolutionary change). The result is that languages tend to spontaneously conform to fit human cognitive limitations and predispositions (especially those of young children's brains), and human brain functions have been shaped by the ecologically atypical features of language. This evolutionary asymmetry means that although human brain structure reflects some of the unprecedented features of language, only the most invariant sensory, motor, attentional, and mnemonic demands imposed by symbolic language will have left their marks on brain structure. In general, we should expect to find that the neuroanatomical tweaks that have made our brains language-ready are not necessarily the most profound attributes of humanness. Instead, it is the way that these biological tweaks have enabled an unprecedented semiotic dynamic to recruit otherwise unmodified brain systems to perform unprecedented functions that contributes most to our distinctiveness.

TWO HUMAN-UNIQUE MENTAL PREDISPOSITIONS

There are both significant differences in brain function that can be understood as adaptations to symbolic and linguistic demands, and also mental functions that have been fundamentally altered by the symbolic processes that were thereby made possible. In the first category are enhancements of working memory and vocal abilities that have resulted from evolved changes in brain structures, such as the relative expansion of prefrontal cortex and the invasion of the brainstem laryngeal motor nucleus (the nucleus ambiguous) by cortical projections, respectively. In the second category are radical functional changes in the ways we use the brain's two major memory consolidation systems and unprecedented reorganization of emotional systems, both due to the novel ways that symbolic processes reuse older mammalian brain functions. In the remaining sections of this chapter, I will focus on the latter type of change and merely assume the former (which I have focused on in many prior publications, e.g., *The Symbolic Species*, 2012).[5]

One of the most profound distinguishing features of humanness is our distinctive form of memory. And yet, this is not itself a consequence of any evolutionary change in the neuroanatomical substrates supporting the major mnemonic functions. In all species of mammals, learning depends on one or the other of two distinctively different mechanisms for establishing long-term stable memories, each depending on quite distinct neural substrates. The acquisition of skills is accomplished by repetition of an activity, which progressively improves precision and efficiency of the activity, while increasing its automaticity and reducing the need for consciously monitoring production. It is generally assumed that repetition progressively strengthens some complex synaptic pathways and weakens others. Thus retention and recall of this information is facilitated by the way that the signaling has become canalized by redundant synaptic strengthening. Since skill learning is particularly important for motor systems (even though also necessarily coupled with sensory feedback), the major brain systems involved mostly involve a frontal cortex to basal ganglia to thalamus to cortex loop, and a similar cortex to cerebellum to thalamus to cortex loop. Damage to structures or connections within these loops significantly impairs this sort of memory formation. This mnemonic strategy is often called procedural memory for these reasons. It exemplifies the general principle that mnemonic strength and accessibility is a function of statistical redundancy.

In contrast, it is also critical to be able to store and retrieve experiential information in contexts where repetition is not possible. What has been called episodic or declarative memory is memory for events or episodes that occurred once, uniquely, are not repeated, and involve little in the way of repeatable actions. This requires the generation of redundancy of a different sort: redundancy of associations. This creates memory traces for singular experiences by correlations between features. Thus when we try to recall a specific experience from our past, it is generally necessary to triangulate to it using correlated associations, involving dates, places, typical social frames, and so forth. This form of mnemonic redundancy is formally orthogonal to the logic of procedural memory, and is thereby supported by quite distinct neural substrates. Thus, episodic memory is generated by neural circuits linking sensory cortices with the hippocampus, and hippocampal damage significantly impairs the ability to consolidate new episodic memories.

Because of this functional segregation of these mnemonic systems, language can play an interesting mediating role. Indeed, it has become the foundation of an unprecedented new form of memory. Early in the process of language acquisition articulatory and syntactical combinatorial skills are acquired procedurally. In contrast, the symbolic reference that constitutes word meanings and their penumbra of semantic and experiential values are necessarily acquired by associative processes. Because of this dualistic use of mnemonic systems, language enables each mnemonic system to reciprocally cue the other. Narrative memory is the result. It forms the basis for promising, reasoning, theorizing, creating our identities, histories, politics, and art. Essentially, every form of socially maintained pragmatic knowledge, from religious belief to technology, is built from a growing matrix of narrative forms.

So although the neural substrates supporting these distinct mnemonic systems have not been fundamentally altered in human evolution from the ancestral primate condition, human cognition has been radically restructured by this novel mnemonic capability. The effects of this on the nature of human identity, agency, and social organization, as well as on the capacity of social groups to acquire and preserve complex knowledge over time, cannot be overestimated.

Symbolic capabilities have also, incidentally, produced uniquely human forms of emotional experience. Like the unique functional synergies that have reorganized the way that ancient mnemonic systems can be used, symbolic capacities have similarly reorganized the functions of the emotional systems of the brain. This has given rise to a whole class of human-unique emotional capacities.[6] These might more accurately be described as symbolically modulated emotional relationships that also are realized by neuronal systems that we humans share with most other mammals.

Because symbols represent their content indirectly and without sharing attributes or direct correlations with the thoughts they convey, the salience and intensity of their emotional correlates are also substantially reduced. This enables symbols to be combined and juxtaposed in many more diverse ways than other sorts of signs. This combinatorial freedom can lead to the expression of emotional interactions that could not otherwise occur. Because the correlated emotions aroused by symbol combinations are of low intensity, they, too, are more easily manipulated and combined in novel ways. So it is not that we have evolved novel neurological systems for emotional expression, but that these processes can be set into novel synergistic and antagonistic and complementary combinations that would be very unlikely to occur in the absence of symbolic processes.

I argue that such complex emotions as awe, nostalgia, righteous indignation, aesthetic appreciation, humor, irony, eureka, and so on, are all the result of the superpositions and interactions of emotional processes due to symbolic manipulation. All involve unusual juxtapositions of more basic emotional dynamics, likely activated differently in the distributed structures responsible for emotion, including differences in homologous structures in the two hemispheres (such as the amygdala and nucleus accumbens and their respective cortical correlates). Not only can this involve the separate circuits that handle different arousal and hedonic states, but the bilaterality of these systems may also allow novel combinatorial interactions of otherwise mutually incompatible emotional dynamics in response to the flexibility of symbolic manipulation. This may help to explain the human fascination with activities that symbolically tweak our emotions in unusual and surprising ways, such as in art, music, dance, ritual and so on.

So, the virtual world we live in encompasses every aspect of our experience and sense of self. Far from merely modifying human communication and intellectual capacity, the evolution of our symbolic capacity has modified nearly every aspect of what makes us human. The result is that humanness is inseparably symbolic and biological. We are a symbolic species.

TWO

Religion in Human and Cosmic Evolution: Whitehead's Alternative Vision

Matthew T. Segall

THIS CHAPTER EXPLORES the evolutionary origins of human religion. As many postcolonial anthropologists have argued, "religion" is a highly contested term that cannot be unproblematically deployed as a transhistorical and universal catch-all category. Although I have chosen to use the word, I agree with this problematization of *a priori* definitions of religion, which all too often blur our perception of the multifaceted richness of human spiritual expression by forcing it to submit to the discursive categories of modern scientific and sociological methodologies. I include the term "spiritual" here to indicate that by "religion" I do not just mean a set of clearly articulated dogmas in which one believes with certainty, but an open-ended and experientially grounded orientation to the mysteries of life and death. Religion and the spirituality, at a core level, are more than can be captured by any fixed definition. They are interrelated dimensions of an ongoing, cosmologically emergent activity, not simply a set of verbally professed beliefs. Instead of trying to explain religion by reducing it to the favored terms of modern biology, psychology, or sociology, this chapter proceeds by attempting to let religious phenomena reveal themselves by situating them within the evolutionary account offered by sociologist Robert Bellah and the cosmological scheme provided by philosopher Alfred North Whitehead.

Inquiring into the origins of religion—and connecting those origins to the evolutionary emergence of our species—is necessarily to step beyond the bounds of strictly empirical or positivist science and into the domain of mythmaking. I approach my topic through what Bellah, after Eric Voegelin, called *mythospeculation*, a method somewhere between theory and story, incorporating elements of each. It is important to be upfront about this, since it does a disservice to the phenomena in question to pretend that what is essential to them could be accessed in an impersonal or objective way. Religion, now and in the past, has more to do with *personal and interpersonal transformation* than with neutral descriptions of factual states of affairs.[1] Inquiring into religion's nature can never be a dispassionate affair decidable by mathematical proof or laboratory testing. At the same time, human religious concerns and values *are themselves matters of fact* that have arisen and continue to arise in the course of cosmic evolution, at least in its human mode. As such, religious concerns require interpretation within any adequate cosmological scheme.

Even the most sober-minded materialistic scientists, whenever they offer evolutionary accounts of the origins of our species or of our universe, inevitably become mythmakers. Bellah makes this quite clear when, in the early chapters of *Religion in Human Evolution* (2011), he examines the popular works of scientific luminaries like Steven Weinberg, Richard Dawkins, and Jacques Monod. It became even clearer to me when I watched the philosopher and author of *The Atheist's Guide to Reality* (2011) Alex Rosenberg, during a recent conference presentation, introduce Charles Darwin and Lord Kelvin as "old testament fathers" and describe images of a leaf insect, a double helix DNA molecule, and a chamber full of gas particles as "iconography"—that is, religious icons whose contemplation is supposed to convert us to the indisputable laws they express.[2] Each of these supposedly scientific thinkers ends up offering their own physical or biological sermon, pretending all the while to have achieved some sort of heroic post-religious and therefore purely scientific rationality. The implication is that they are mature, enlightened adults while the rest of us are cowardly children afraid to accept the meaninglessness of our own existence, terrified of the fact that we are, as Monod put it, "[gypsies living] on the edges of an alien world."[3]

In contrast to these scientific thinkers engaged in what Whitehead referred to as "heroic feats of explaining away,"[4] my approach, building on Whitehead and Bellah, is motivated by the search for a cosmological reconciliation between scientific theorization and religious mythopoeia. I hope to show that the forced choice between religion and science is a false one, and that the emergence of an ecological civilization depends upon our species' ability to enact a cosmological vision that does justice to both scientific facts and religious values, and that recognizes the various ways facts and values inform one another.

One of the most well-known attempts to explain away the phenomenon of religion is the philosopher Daniel Dennett's book *Breaking the Spell: Religion as a Natural Phenomenon* (2006). He begins his book by comparing religion to *Dicrocelium dendriticum* (lancet fluke), a tiny manipulative parasite that infects the brains of ants, compelling them to climb to the top of the nearest blade of grass so as to get themselves eaten by a cow, thereby transporting their fungal stowaways into the nutrient rich environment necessary for the completion of the latter's reproductive cycle. Religion is explained, not as a *genetic* parasite, but, building on Richard Dawkins' well-known but scientifically discredited[5] meme theory, as a *memetic* parasite, a sort of mind disease. By analogizing cultural evolution to the blind process of natural selection, even mind is explained away as mere mimicry: *Monkey see, monkey do.* So-called "religious memes" are said to spread and survive today not because past peoples found them deeply meaningful and transformative but because they have succeeded in their "competition for rehearsal space in the brain"[6] by getting copies of themselves made.

To be fair to Dennett, his book is less an attempt to provide the definitive explanation for the evolution of religion than it is an argument that religion ought to be studied scientifically as a natural phenomenon. He admits that the memetic theory he puts forward is probably wrong, but at least, he says, it gives others something to fix. Fair enough. Following thinkers like Bellah and Whitehead, I am sympathetic to the call for a naturalization of religion, for a scientific study of its emergence out of a wider biological and cosmological context. But of course, it all depends what we mean by "science" and what we mean by "nature." There is more than one kind of naturalism.

The problem is that approaches like Dennett's to the evolutionary emergence of religion presuppose what Whitehead's philosophy of organism so passionately protests against: *the bifurcation of nature.*[7] For Dennett, to count as a scientific explanation, the cultural meanings of religion must be accounted for in terms of the natural mechanisms of his reductionistic view of biology. All the seemingly intrinsic values of our human existence must once have been of merely instrumental survival value, otherwise they could not have been preserved by the Darwinian mechanism of natural selection. All seemingly intrinsic value is thus explained away as a mere "psychic addition" to what is really the purposeless exchange of genetic or memetic material from brain to brain across the generations.

The contrast between such reductionistic biological accounts of religion, and Bellah's and Whitehead's integral, cosmological approaches, ,could not be starker. Dennett mentions and even praises William James's radically empiricist approach to religious experience (a major influence on Whitehead), only to dismiss it as inadequate for his own, more reductionistic purposes. Dennett instead trades in James's psychological view for what he describes as a wide-angle biological and social (or sociobiological) lens. For Bellah, and especially Whitehead, while biology, psychology, and sociology each have important contributions to make to the study of religion, in the end the proper lens to take must be maximally inclusive: human religious expression must be understood in the broadest context we are capable of imagining, namely, the cosmological.

"Cosmology," writes Whitehead, "is the effort to frame a scheme of the general facts of this epoch, of the general character of the present stage of this universe. The cosmological scheme should present the genus, for which the special schemes of the sciences are the species."[8] He goes on: "A cosmology should above all things be adequate. It should not confine itself to the categoreal notions of one science, and explain away everything which will not fit in. Its business is not to refute experience, but to find the most general interpretive system."[9]

So long as our view of nature falls victim to the fallacy of bifurcation, reductionistic explanatory strategies like Dennett's will continue to handicap scientific investigation into the evolutionary emergence of religion. Instead of trying to explain away religious phenomena as the accidental result of blind biological forces, we can more coherently approach it as a genuine flowering of the universe we find ourselves living within. Treating religion scientifically requires coming to view it not as an improbable anomaly, but as a natural expression of cosmogenesis in its human mode. Human religious experience, in other words, must count as part of the legitimate data to be included in any adequate account of this universe. To treat religion naturalistically, we need not explain it away as epiphenomenal. We can instead inquire into the cosmic conditions of its possibility. From the perspective of Whitehead's cosmological scheme, the history of the human species' religious experience "consists of a certain widespread direct apprehension of a character exemplified in the actual universe."[10]

Stated in more general terms, instead of following the typical reductionistic logic of evolutionary explanation that seeks to make life and mind mere epiphenomena accidentally emergent from what remains in reality a dead material universe, we can adopt the alternative, no less naturalistic Whiteheadian approach. "[Humankind] has gradually developed from the lowliest forms of life, and must therefore be explained in terms applicable to all such forms," admits Whitehead. "But why," he continues, "why construe the later forms by analogy to the earlier forms. Why not reverse the process?"[11] That is, why not give up the polemical desire to explain away the more complex by reducing it to the less complex by recognizing that, if phenomena like life and mind (and with them, human religiosity) are present in today's universe, they must have in some sense been prefigured from the beginning.

"In the course of evolution," Whitehead asks, "why should the trend have arrived at [humanity], if [our] mental activities ... remain without influence on [our] bodily actions?" In other words, the question we should ask ourselves is *what is this universe such that something like human organisms with their religious mentalities are possible?* Whitehead's answer is that "some lowly, diffused form of the operations of [mentality] constitute the vast diffused counter-agency by which the material cosmos comes into being."[12] This "counter-agency" confronts the otherwise entropic tendency of the physical universe, a tendency Whitehead has no interest in denying. Much of the cosmos, including the Sun that feeds all life on our planet, he readily admits, is decaying and will eventually return to chaos. He invokes a counter-agency only out of explanatory necessity, since the mere mechanics of efficient causality cannot account for the current highly organized state of the universe, for the fact that a star like the Sun feeding a living planet like the Earth should have been energetically possible. Physicists now understand that, far from equilibrium, systems are not in fact disobeying the 2nd law of thermodynamics, but more efficiently realizing it. But why must we emphasize entropy as the sole causal tendency, given that physicists now also understand our universe to be self-organizing at every scale? Why not also identify *centropy*, the tendency of our universe to organize itself into ever-more complex forms or centers of agency? Alongside efficient causality, formal and final causality are also evident in the creative urge of the universe toward as yet unactualized possibilities of self-organization. If we deny a cosmic ground to agency, purposiveness, and value, metaphysical consistency requires the absurdity that we deny these in ourselves, as well. For we are the children of this universe. Whitehead defines the advanced cosmological stage of religion as "the wider conscious reaction of [humans] to the universe in which they find themselves."[13] Following Whitehead's reversal of the usual logic of evolutionary explanation, we can recognize the emergence of religion in human beings as evidence that something more than blind chance and inexplicably imposed physical laws are at work—or, as we'll see—at *play* in the evolution of our universe.

Bellah, like Whitehead, grounds his account of the emergence of religion in the broadest possible context by situating human evolution within so-called "Big History." In his book, *Religion in Human Evolution,* Bellah spends the first 40 pages of the second chapter, "Religion and Evolution," laying out the course of cosmogenesis from the first few seconds after the Big Bang, through the formation of galaxies and stars, to the solidification of the Earth, to the appearance of the first single-celled prokaryotes, to eukaryotes, metazoans, reptiles, mammals, primates, and finally Homo sapiens. He is less confident than Whitehead when it comes to attributing some "metaphysical direction" to the overall arc of the evolutionary process. He does, however, approvingly reference a comment in *The Origin of Species* where Darwin admits that "a little dose...of judgement or reason often comes into play, even in animals very low in the scale of nature."[14] Purpose does seem to operate, then, at least at the scale of individual living beings. In contrast to Dennett's mechanical gene-centric view, Bellah's, like Darwin's, is certainly an organism-oriented understanding of biology. But it is not yet a full-fledged *ontology* of organism like Whitehead's. More on this in a moment.

Although Bellah recognizes important distinctions that make humans unique among other members of the animal kingdom, even reproducing Terrence W Deacon's statement that our species represents an entirely new phylum, he nonetheless dwells at length on the many preexisting mammalian capacities that prepared the way for us, including extended parental care, empathy and shared attention, ethical relationality (including ritualized aggression and mating), and, most significantly, the capacity for play. Play becomes especially prominent in young mammals because of the "relaxed field" provided by prolonged empathic parental care. This period extends even more as evolution draws nearer to Homo sapiens, who are born exceptionally premature and remain in the childhood phase longer than any other species. Play is not initially a functional capacity that might be selected for by the normal Darwinian mechanisms. Play is evidently engaged in purely for its own sake: it is an end in itself. Play has nothing directly to do with sexual reproduction or eating (though it may be erotic and enjoyable), nor can we play while fleeing or fighting for our lives. This is not to say that play may not become functional later on. Bellah cites numerous ethologists who describe the way bouts of playfulness in some primate species leads to the neutralization of hierarchies and physical inequalities among play partners, such that a sort of proto-justice appears to emerge. More than any other animal behavior, play requires the capacity, not only for shared attention, but for shared intention. Shared attention and intention are the precondition for any form of empathy or sociality.

Most significantly for the purposes of this essay, Bellah posits that early hominids developed the first ritual activities out of complexified forms of mammalian play. The source of the complexification was the ramping up of empathic sociality among humans, eventuating in what Bellah (quoting Sarah Hrdy) calls "emotional modernity."[15] Human minds, due to their tendency to play ever-more intimately, have become uniquely empowered (and sometimes possessed) by symbolism—the ability of words and images to bind us to certain sociopolitical realities, realities we cocreate in concert with deep cosmic and biotic patterns through ritual enactments of myth. This power of symbolic binding transforms ritual play into religion. It is important in this context to admit, as Whitehead reminds us, that "we should not be obsessed by the idea of [religion's] necessary goodness. This is a dangerous delusion."[16] Despite the fact that religious symbolic consciousness was born out of our unprecedented capacity for social intimacy, once it has emerged, it can also detach us from one another just as readily, generating the worst kind of in-group/out-group discrimination and violence, and, as has become more apparent in the modern, industrial era, symbolic consciousness also has the power to produce civilizational myths that are entirely detached from the ecological context of the living planet that sustains us.

While symbolic consciousness may be the *flower* of religion, it grows from seeds planted in the soil of collective ritual play. Religion is not primarily a matter of individual belief: it is rather something we are and do together, not just with other human beings, but with the community of life on Earth and the broader cosmic community. The essential thing about religious life is not the mindless repetition of dogmatic creeds, but sincerity in its engagement with symbolic forms of ritual play. According to Whitehead, a religious symbol "[has] the effect of transforming character when [it is] sincerely held and vividly apprehended."[17] Early rituals, we can speculate based on the archeological evidence, emerged out of collective celebration involving song and dance. Most probably, these celebrations were in tune with diurnal, lunar, and seasonal rhythms. The earliest religious rituals were cosmologically embedded celebrations of the cycles of life, death, and rebirth. These ritual celebrations were not based on beliefs in supernatural beings, but on deep perception of, and desire to participate in, the rhythms animating the plants and animals of the Earth and the shining orbs in the sky. As Jason Kelly puts it elsewhere in this volume, it could be argued that, at the root of all religious expression lies the desire to realize "cosmic consciousness," that is, the desire "to become 'one' with the forces of nature."[18] The human being's religious impulse, growing out of ritual play, is to "recreate" the harmonies of these cosmic forces in symbolic form, to refashion them into myths for the guidance of our civilized societies. Only very recently in the history of our species have these ritualized symbolic enactments become detached from their encompassing cosmic and biotic rhythms. Our modern myths have become too anthropocentric. We have immersed ourselves in a symbolic system that is radically out of tune with our ecological context.

Bellah's argument draws extensively on the cultural historian Johan Huizinga's book *Homo Ludens: A Study of the Play-Element in Culture* (1938). Huizinga argues that "in the form and function of play... [humanity's] consciousness that it is embedded in a sacred order of things finds its first, highest, and holiest expression."[19] Rooting the emergence of religion in ritual play short-circuits any attempt to explain religion in terms of biological utility, since by definition play is not about working as a means to the ultimate end of survival, but about sheer enjoyment as an end in itself. Further, because of the important role of play in the evolution of our species, and because it depends on shared attention/intention and basic ethical relationality, it provides clear evidence against Dennett's view that organisms are just mimicry machines. "In acknowledging play," writes Huizinga, "you acknowledge mind, for whatever else play is, it is not matter." "Even in the animal world," Huizinga continues, "[play] bursts the bounds of the physically existent. From the point of view of a world wholly determined by blind forces, play would be altogether superfluous. Play only becomes possible, thinkable... when an influx of *mind* breaks down the absolute determinism of the cosmos."[20]

Huizinga here almost slips into Whitehead's fallacy of bifurcation by reifying the difference between mind and matter. Elsewhere Huizinga asks "would it be too absurd to assign a place [to play] *outside* the purely physiological?" I would say yes, it would be absurd, or at least incoherent, to suppose the playfulness of mind-bearing organisms somehow exists in a realm that is ontologically separated from their physiology. The physiological need not be equated with the mechanical. As we will see, Whitehead's panexperientialist ontology allows us to imagine other possibilities.

Even though I'm critical of Huizinga's slippage toward bifurcation due to his tendency to reify culture and mind as entirely "outside" of and set apart from mere "nature," I still acknowledge and gladly amplify his other, underemphasized but no less profound intuition, that the efficacious reality of play in human and nonhuman lifeforms entails that we inhabit an intelligent, sensitive, sometimes violent and sometimes playful universe, not a dull, deaf, and dumb one. As I suggest below, I have similarly mixed feelings about the residue of bifurcation in Bellah's more culture-centric and phenomenological approach to religion.

We might also describe ritual as *serious* play (following Huizinga who points out that the opposite of play is not seriousness, but *work*). That animals should engage in play behavior is already a sign that reductionistic accounts of biological evolution miss something when they ignore organismic agency by focusing exclusively on the struggle for existence and fitness to a preexisting environment. Life, as Whitehead also knew, is not just about mere survival. The life seeks more than mere survival: it seeks to thrive, to "live well, and to live better."[21] If survival were the name of the game, matter would have done better to remain in rock form, for compared to million-year-old minerals, life is deficient in survival value.

Whitehead, like Bellah and Huizinga, also roots religious behavior in ritual forms of play.[22] Both he and Bellah offer strikingly similar accounts of the stages of religion's evolutionary emergence (see Figure 1 on page 28). Both Whitehead and Bellah acknowledge that ritual is widespread among mammals. Early humans were no different, but because of their increasing emotional and cognitive sensitivity they began to recognize that certain emotional states, enjoyable for their own sake, apart from the needs of biological survival, could be reliably reproduced through collective ritual enactment. Only later, once the capacity for symbolism emerged, were mythic beliefs articulated as expressions of the purpose of ritual practices and their attendant emotional qualities. Myths then contributed recursively to the intensification of the emotional qualities. Notice that the arrows in the diagram point both ways, which is meant to prevent us from thinking that the emergence of a new stage means the prior stage is forgotten or transcended. Early stages are still present with, and necessary

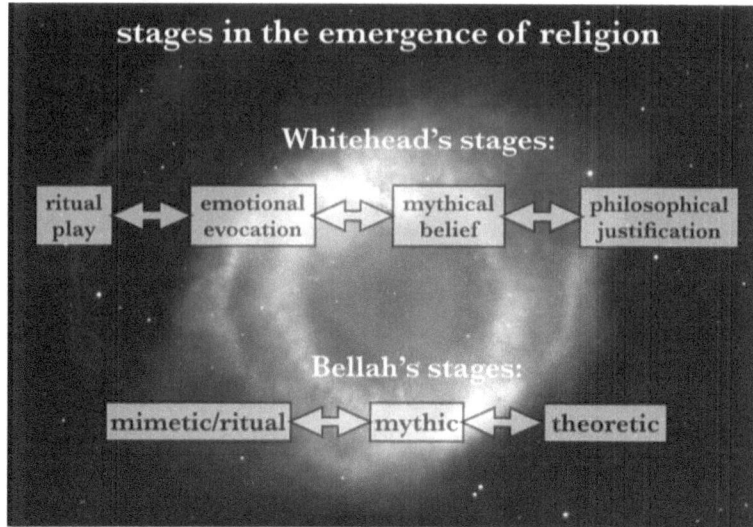

Figure 1 Stages in the Emergence of Religion

for, the expression of later stages. This is true even with the latest stage of rational, philosophical, or theoretical reflection upon religious rituals and myths. Religion of the theoretic or rational type (the sort we are most familiar with today) grows out of, and remains dependent upon nonrational forms of mythic speech and ritual play. Again, an adequate account of the emergence of religion in human evolution makes it clear that it is not primarily about *what* one believes, but about *who one is* and *what one does*. The fundamentalisms of our late modern age, whether atheist or creationist, tend to neglect the ritual and mythical dimensions of religious life. Instead they focus almost exclusively on the cognitive components of belief systems, which are often only the dead products excreted by a more primary living process of cosmic participation. Explicitly stated beliefs are the most superficial aspect of human religion. Given Whitehead's non-bifurcated and re-enchanted cosmological scheme, the myths generated by ritually induced emotional upwelling need not be dismissed as childish fairy tales, but can be understood to be the archetypal energies of the cosmos itself erupting into human symbolic consciousness.[23]

Bellah describes ritual play as an experiential opening transporting us into a non-ordinary reality transcending the everyday world of "work" or mere survival. Bellah's understanding of religious experience as one among a variety of cultural realities (differing from that of science, aesthetics, politics, and so on) is drawn largely from the cultural phenomenology of Clifford Geertz and Alfred Schütz. This sort of phenomenological approach provides a helpful critique of, and alternative to, more scientistic explanations by allowing us to examine religion on its own terms. Indeed, as I describe below, Bellah's use of Martin Buber's theological phenomenology provides crucial insight into the nature of religious concerns. But because, in general, phenomenological approaches, especially those with a cultural and symbolic focus, leave the question of the *cosmological* basis of religious experience unanswered, if not also unasked, I argue that a Whiteheadian speculative supplement is necessary. Taking a phenomenological look at religious experience by bracketing other cultural enactments of reality risks leaving the bifurcation of nature from culture intact. Whitehead's scheme allows us to grant the validity of multiple cultural realities while also acknowledging human culture's continuity with the nonhuman cosmos. How it does so will become clearer as I conclude this essay, but before jumping ahead I must continue unpacking Bellah's important claim that ritual play (and the religious experiences it is associated with) transcends the everyday world of work. The idea is not to transcend work entirely, which would be impossible, but to recognize its relativity in relation to all the other experiential realities in which we participate. A certain degree of work will always be necessary for survival, but the question remains: what we are to survive for, if not to engage in ever-more ingenious forms of play (and here "play," following Huizinga, should be taken in its widest sense as the basis for all sociocultural activity), then for what? And what does it mean that ritualized play, and the spiritual efflorescence it generates, is at the historical origin and remains the existential core of our cultural lives?

One way to apply Bellah's theory is to consider what it tells us about the history of work, in particular as it relates to the shift in socioeconomic organization represented by the agricultural revolution. Göbekli Tepe, an enormous and elaborately decorated 12,000-year-old temple structure uncovered in the 1990s by archeologists in Turkey, provides us with a counterexample to the standard, techno-centric account of human evolution. As the standard account goes, human beings needed to technologically secure their basic survival needs by domesticating plants and animals before the supposedly superfluous activities of ritual, art, and religion (all closely related for archaic consciousness) could flourish. The existence of Göbekli Tepe suggests, instead, that these cultural activities predated the shift to the agricultural mode of production. Evidence at the site shows conclusively that the people who built this temple were hunter-gatherers. It does not seem such a stretch to suggest in light of this site's age that the need for stable religious expression made the labor-intensive shift to agriculture more worthwhile than it otherwise would have been for hunter-gatherers, the "original affluent society," as the anthropologist Marshall Sahlins has argued.[24] The tremendous amount of detailed planning and hard work required to construct such a temple—a structure we may suppose produced for the people who constructed it a ritually enacted, relaxed field of spiritual and artistic play—makes clear that no necessary separation exists between the serious and the playful. Human beings are quite willing to work harder to secure time and space for more elaborate forms of play. Not only religion, but science and art too, are born out of our innate playfulness. Humans are not the only beings who play, but surely we have taken play more seriously than any being before us.

This understanding of the origins of religion (and culture more generally) in ritualized play provides a powerful critique of the political economy guiding our contemporary civilization, for which work has become an end in itself, and for which play, when we find the time for it, has little connection to the rhythms of the Earth and wider cosmos in which we are embedded. The question remains: Are we here to toil extracting Earth's resources, competing with one another for more money to consume more products, or are we here to ritually participate in ever-renewing cycles of cosmic creativity?

Part of what makes so many scientific materialists averse to accounts of the evolution of religion like those of Whitehead and Bellah is that they seem at first to be both too anthropocentric and too anthropomorphic. When Whitehead claims that photons, protons, electrons, stars, and galaxies are species of organism possessed of feelings and desires, and that their ecological evolution is analogous to that of bacteria, plants, and animals, is he not just projecting human or at best vital capacities onto a dead, inanimate collection of objects? *Only if we are unwilling to reconsider the incoherence of modern science's bifurcation of nature.* What if the scientific attitude of austere objectivity makes the scientist constitutionally tone deaf to the erotic pulse of the universe? Overcoming the incoherence of the bifurcation of nature requires a new scientific outlook, since the narrow materialist version of science makes it impossible to understand how life and consciousness (not to mention religious expression) could be a part of this universe. We are left having to claim they are astronomically improbable accidents, which is the exact opposite of an adequate scientific explanation. What if instead of turning our own existence into an absurdity we look again at the universe and ask: *What is this universe such that something like human organisms with their religious mentalities are possible?*

This is not to center the universe on the human, or to make the universe in the image of the human, it is only to admit the evident fact that we are the children of this cosmos. For better and for worse, the space-time of this universe is our parental unit. We are not an accidental appearance in this world; we are what the universe has come to be doing here and now, an anthropic amplification of its innately prefigured potentiality. I have metaphorically referred to our species as the children of this universe several times in this chapter, which would seem to play right into the hands of scientific materialist atheists, who tend to rhetorically position themselves as the only adults in the room. But I have used these metaphors deliberately because I think the story of heroic maturation from naïveté into nihilism championed by Monod, Dawkins, and Dennett, et al., contradicts the evolutionary evidence that what makes our species so unique is precisely our "childishness," that is, our neoteny and propensity to play. Maturation and adulthood need not be defined by the acceptance of cosmological meaninglessness. It is precisely this attitude that has led more and more members of our civilization to resign themselves to a life of toil to accumulate the only remaining value "adults" are allowed to believe in: money (and maybe power, too). "Truth" may be of value to the scientific materialists I have mentioned, but it seems to me that when they rhapsodize about their desire to understand the universe they almost always fail to hold their own value-laden view of truth to the same skeptical standard they hold those with (explicitly) religious views of truth to. *If* we are to allow biological, psychological, or sociological explanations for religious truth-values, then we must also allow such explanations for scientific truth-values.

Bellah is not as metaphysically confident as Whitehead about the cosmic extent of meaning or the centropic tendency of evolution. But he is by no means a cosmic pessimist like Dennett, Monod, or Rosenberg. Bellah takes his stand not on an ambitious metaphysical cosmology, but on the phenomenological theology of Martin Buber (thereby potentially helping him overcome the residue of bifurcation resulting from his reliance on Geertz and Schütz's more culturally focused, and so ontologically underdetermined approaches). Buber distinguished two fundamental ways of relating to reality: 1) the I-It relation, which objectifies the world into dead things to be manipulated, and 2) the I-You[25] relation, which perceives the world as full of subjectivities, and as itself a subject (i.e., God, the "eternal You").

Building on Buber, Bellah argues that it is not at all surprising that for a "supersocial" species like us, an "I-You relation would at the highest level of meaning trump the I-It relation." He continues:

> To put it bluntly, there is a deep human need—based on 200 million years of the necessity of parental care for survival and at least 250,000 years of very extended adult protection and care of children, so that, among other things, those children can spend a lot of time in play—to think of the universe, to see the largest world one is capable of imagining, as personal.[26]

Understanding how religion could have emerged from mammalian play requires shifting from the I-It to the I-You mode of relation. "In the observation of play," writes Bellah, "and even more clearly in actually playing with an animal, it is almost impossible not to have an I-You relation, which arouses suspicions that one is not really doing science."[27] The I-It relation leads the scientific materialist to a view of evolving organisms as passive machines, rather than creative actors. Grasping the creative, purposeful, playful dimension of organic life requires the adoption of a more participatory I-You relation to evolution, which is what Whitehead invites us to do when he reverses the typical logic of evolutionary explanation. In contrast, Dennett's I-It approach is predicated upon the idea that the best way to study the evolution of religion is to imagine we are aliens from another planet trying to gain a view of it "from the outside," as it were. To approach human religion from such an alienated perspective is to seriously handicap the pursuit of a naturalistic account of its evolutionary emergence. If we want an account of religion's emergence that is *immanent* to cosmogenesis and avoids the undue imposition of otherworldly transcendence, then we are going to need to study religious experience from the *inside out*.

"The final principle of religion," writes Whitehead,

> is that there is a wisdom in the nature of things, from which flow our direction of practice, and our possibility of the theoretical analysis of fact . . . Religion insists that the world is a mutually adjusted disposition of things, issuing in a value for its own sake. This is the very point that science is always forgetting.[28]

Science deals with the facts, but some scientists, in their perhaps somewhat adolescent, hubristic rush to overthrow the religious social matrix out of which science emerged a few hundred years ago, have neglected to include the values of the universe alongside the facts, or rather, to include these values as among the facts. "We have no right," writes Whitehead, "to deface the value experience which is the very essence of the universe."[29] For what is a fact, metaphysically speaking? Whitehead's non-bifurcated image of nature is a rejection of the fallacy of "vacuous actuality"—a rejection of the idea, in other words, that facts can exist independently of experiential values. To be actual, to be a fact, for Whitehead, means to experientially enjoy existence as an end in itself, to value oneself as an actuality and to be valued by other actualities. Without the value-experience of human and nonhuman organisms, "there is nothing, nothing, nothing, bare nothingness."[30]

Whitehead's cosmology is an invitation to move beyond the modern bifurcation separating nature from culture, fact from value, and mechanism from meaning. Moving beyond the bifurcation of nature to grasp the cosmological significance of religion, and the religious significance of cosmology, will require reevaluating metaphysical assumptions that have been woven into the very fabric of the scientific worldview for hundreds of years. The originators of this worldview, the original mythmakers responsible for initiating the Scientific Revolution, conceived the universe as a machine and imagined God as its transcendent designer. Though they differ in the details, this was the imaginative background informing the thoughts of Newton and Descartes. Nowadays, scientific materialists no longer have any need for the "God hypothesis" as Laplace famously called it, but the imaginative background informing their ideas remains the same. The universe is still to be understood by analogy to a machine, only now it has become a purposeless machine. Understanding this cosmic machine requires purifying our perspective of any hint of emotion, value, or aesthetic appreciation, since these merely subjective qualities are thought only to contaminate an impartial view of reality. Whitehead's cosmological scheme provides an alternative.

"The metaphysical doctrine, here expounded," he writes in the final pages of *Religion in the Making*,

> finds the foundations of the world in the aesthetic experience, rather than—as with Kant [and many contemporary scientific materialists]—in the cognitive and conceptual experience. All order is therefore aesthetic order ... The actual world is the outcome

of the aesthetic order, and the aesthetic order is derived from the immanence of God.[31]

To conclude this chapter, I would like to draw a parallel between Whitehead's aesthetic ontology and Huizinga's understanding of play, a parallel generative of a series of fertile questions worth considering. Huizinga locates play within the field of aesthetics, and suggests that play is inherently generative of order. "Play," he writes, "has a tendency to be beautiful."[32] Huizinga goes on, in Whiteheadian fashion, to describe ritual acts of play as cosmic happenings that are continuous with natural processes.

Would it be too absurd, following Whitehead's rejection of the bifurcation of nature in favor of an aesthetic ontology, to assign a place to play *within the evolution of the universe itself*? Might we come to understand the whole of the cosmos at every level of its self-organization as an expression of divine play? Might Blake have been right that "energy is eternal delight"? Instead of God the disincarnate transcendent designer of a clockwork universe, or a meaningless machine-world running down toward heat death, might we interpret the scientific evidence otherwise? Might it be, as Whitehead suggests, that "the world lives by its incarnation of God in itself," that "every event on its finer side introduces God into the world," that "every act leaves the world with a deeper or a fainter impress of God"?[33] For those with an allergy to the "G" word, remember that Whitehead's philosophical intervention into traditional theology aimed to transform the transcendent God of "coercive forces wielding the thunder" into the creaturely God of persuasion, "which slowly and in quietness [operates] by love."[34] The ultimate religious theme in Whitehead's cosmology is this divine Eros, the counter-agency that saves the world from decaying into irrelevance by luring organisms toward more creative forms of organization. Whitehead's God is not a big boss in the sky who designs and determines everything, but the poet of the world—indeed, the *tragic* poet of the world—who through aesthetic sensitivity beckons all beings toward the highest beauty that is possible for them given the limitations of their finite situations. *Beauty is the teleology of the universe.* This, at least, is Whitehead's alternative cosmological interpretation of the facts and values of the history of human religious expression. Whether or not we seize this alternative vision will determine the future of our civilization, if indeed it is to have one.

THREE

"Are We Human or Are We Dancer?"

René Pikarski

INTRODUCTION: TOWARDS A DYNAMIC IDEA OF THE HUMAN CONDITION IN THE LIGHT OF LIFE AND POWER

THIS CHAPTER[1] will be guided by a question asked in the famous lyrics by the band The Killers: *"Are we human or are we dancer?"*[2] As interesting as the question is in itself, there is also its implicit assumption of a radical distinction: the "or" denotes a discontinuity that tries to separate being human from being a "dancer," that is, a marionette. Possibly, it awakens memories of two very classic notions. On the one hand, one may find what is often subsumed under the notion of a more or less "free" will; for instance, thought as the ability to set purposes and intentions for one's actions—"free will" as an attribute of the human mode of being in the world. On the other hand, there is the notion of an unfree dancer whose actions are entirely determined by foreign arbitrariness and constraints. The *transmission* from a mechanical marionette to the famous comparison between automates and animals is not such a big leap. Whether marionette, automate, or animal, in these ideas, lack of freedom is an attribute of nonhuman modes of being in the world. These old radical discontinuities are often corrected, refined, revised, or completely overwritten by a large amount of modern discourses. The concept of freedom remains, although not necessarily as a discriminating or absolute attribute. Likewise, the question of which attributes mark the difference between human beings and other lifeforms is of great interest. Thus, the anthropological counterpart to the pop music question is *"What is the human being?"* Even though this is often called the Kantian question, the second European Enlightenment is not the only one to give special attention to it; the first, ancient Enlightenment did, too. The question can be found in several ancient discourses, such as Plato's dialogues.

What both movements have in common is something that I call an *unproblematic core*. Indeed, the human is often seen as a problematic living being, but the answer to the question itself—the definition—has throughout time often appeared as unproblematic. One can see this in the following comparison. In Plato's *Alcibiades*, Socrates answers the question "Then whatever is man?" by saying "either man is nothing at all, or if something, he turns out to be nothing else than soul." However, this conclusion is unambiguous only if the definition of the human as man and soul does not obliviate the body, but retains the idea of soul as "the user of the body."[3] Furthermore, in the basic work of the French Enlightenment, the *Encyclopedia* of 1759/1763, the human is categorized as being in both the spiritual and material worlds. According to d'Alembert's famous *Preliminary Discourse*, the human "belongs by virtue of his soul to the spiritual beings and by virtue of his body to the material world."[4] Humanity's ambivalent belonging to both the realms of divinity and the natural world obviously declares its own nature as a highly problematic one, but the definition itself managed to outlast ancient and medieval times, at least until the modern era. However, with the rise of the modern sciences, not only the human itself, but the very definition of the human became a severe problem. A late but significant witness can be found in a commentary by Max Scheler. For him (writing in 1927), the problem with defining the human emerged out of a careless fragmentation of discourses. There is an anthropological, a philosophical, and a scientific definition of the human, but none of them cares about the other.[5] This carelessness refers to the lack of a general idea or to the deficiency that results when the concept of what it means to be human is combined with the continuous differentiation of the life sciences. Read carefully! According to Scheler, creating this idea is a *practice of care*. *The human condition is something we should engage with great concern because the stakes could not be higher*. Even more interesting is how this remark is commented upon by Ernst Cassirer: "But our wealth of facts is not necessarily a wealth of thoughts."[6] In other words, we usually do not *care* about facts—facts are not a matter of care. Creating an idea of the human requires more than a scientifically scattered accumulation of data. It requires more than stating or identifying facts about what the human *is*, facts that are then lifted to elevated levels of formalization and generalization. Thus, the idea of the human must fulfil a science-comprehensive, interdisciplinary, and meaningful role. In this light, with the now problematic question of what the human is, Scheler and Cassirer move very close to Alfred N. Whitehead's later claim of an adequate speculative philosophy, namely, "to frame a coherent logical, necessary system of general ideas in terms of which every element of our experience can be interpreted"[7]—which, of course, includes the human condition. But considering Whitehead as a process thinker, it soon becomes clear that something special should happen with the idea of the human. No matter how it will be postulated, the idea must be revisable; it must be flexible and agile: a dynamic and highly variable idea.

My aim in this chapter is to show how a dynamic idea of the *conditio humana* can be transferred into the concept of critique. A critical stance usually points at something that persists with obstinacy, with the view to change it or at least to create possibilities to crack this persistence and open a less canalized future development. If you look for a *critical* idea of humanity, in terms of a dynamic idea, the human condition is not only a summary of facts about humans, but always reveals a purpose, an imperative, a duty, and an exercise that we must constantly actualize in order *to be* human. Michel Foucault's concept of power will be especially helpful here, as will Henri Bergson's understanding of the notion of evolutive capacities, which are responsible for the coordination of the movements of the living. Both are fertile for the quest for a critical version of the *conditio humana*. Both thinkers and their fundamental concepts of power and life provide ideas that do not state the human condition via the factual "is," but understand it to be a process, an enduring becoming.

This chapter takes a middle position within the current book. If you look back from here, you can see anthropological perspectives on evolutionary processes that try to evaluate the discriminatory and common features between humans and animals in terms of relations of emergence and dependencies, as well as processes of differentiation. Looking forward from here, a dynamic idea of the human becomes essential and more than desirable, provided you consider that humans have immense capacities to cocreate their own being. No matter if one speaks of biological or bio-factual transformations (biological artefacts), all of them would be violated or less minded if we held to a static human condition based only on stated facts. The manifold versions of transhumanism show especially how the open-ended future of human evolution is always, and in its essence, a critical and therefore problematic process.

Here, then, is the idea of what it could mean to undertake a critical definition of the human condition against the backdrop of life and power:

1. First, it means that humans, based on their intelligent and intuitive performances, tend towards a high flexibility, openness, and universality within the configuration of their relations and references towards the world, their environments, others, and themselves. I subsume these abilities under the notion of a *capability for distance from life* and a specific *capability for a closeness to life* that is mediated by that distance.

2. Second, it means that these capabilities and performances are decisive for the spontaneous constitution of human modes of action and the canalization of possibilities for action. These actions generally take place in social frameworks and therefore within power relations. Thereby, one can see how humans, *by developing a self*, can and should bring themselves into a position to critically reflect upon and transform the dependencies that surround and penetrate them. In other words, a human being can and should ongoingly become a subject. The comprehensive topos within which these processes of becoming a subject are embedded is the "care of the self." Finally, Scheler's demand for a *conditio humana* as a practice of care has one possible correspondence to the ways a subject can spontaneously and creatively form its own development, as well as the society within which it lives.

In this context, the common ground between Bergson and Foucault is the notion of freedom. On the one hand, it occurs in a versatile form of human freedom as freedom of action (within power relations). On the other hand, it occurs in the form of freedom with regard to the capabilities needed to perform these spontaneous human actions. This freedom, as previously noted, is actualized in the human intelligence and in its intuitive ways of interacting with the world, its environments, other living beings, and itself.

BIOLOGICAL LIFE AND THE BIO-PHILOSOPHY OF CAPABILITIES

The Human Gap as a Bio-philosophical Approach Path

Several contemporary scientific discourses try to fan out differences between animals and humans in the evolution of intelligence. Thus, to focus on intelligence means to answer the question about discontinuities between animal and human life primarily via mental states and processes. The developmental psychologist Thomas Suddendorf provides a brief overview of this matter. He offers six theses which claim a gap between human and animal intelligence:[8]

1. **Animals communicate—humans have language.** In human symbolization, there is a feature that allows for flexible references, symbols that refer to several objects, even to nonexistent or non-present objects. It allows us to build symbolic hierarchies or generic, abstract concepts, and an open grammar.

2. **Animals may have procedural and semantic memory—but humans have episodic memory.** This memory is a specific mode of thought that allows us to discriminate between past events for the purpose of either initiating a present action or establishing a future perspective.

3. **Animals are capable of specific social cognition—humans have *"theories of mind."*** We have an intuition for the needs, thoughts, and feelings of other persons through not only observing their current state but also by considering their past events and plans for the future.

4. **Animals can be intelligent problem solvers—humans can conceive more creative solutions, for example complex theories.** This creativity not only requires a capacity for abstract thinking and imagination, but also spontaneous intentions, decisions, and actions that are aligned with a relatively distant purpose and aimed at overcoming several obstacles.

5. **Animals are capable of empathy—humans additionally have ethics.** Connected to a flexible mode of thought and language, human normative strategies are broad, universal, and more adaptable. Furthermore, humans have the capacity for moral evaluation and judgements, a capability connected to the ability to imagine

different possible and impossible alternatives.

6. **Certain species pass on traditions—humans have "culture."** For example, a human quality "passed on" is *not* a precise imitation of behaviors and technics of conspecifics. The human way of imitating is more flexible and open to self-instructed improvements.

Principally, a high, *intersubjective situational flexibility* makes us human. This means that we distinguish ourselves from our nearest living relatives, the great apes, by an *"open-ended"* capability for imagination and reflection about different situations, as well as the need for sharing with others the scenarios created by our minds.[9] In the discourses about typical human differences, one rarely comes across well-defined borders with other living creatures. Perhaps the symbolic, recursive nature of human language and some aspects of mental capabilities are the sole exemptions where one can see a relatively strong segregation from other living species. However, what becomes apparent is that the two tendencies of *flexibility* and *universality* are repeatedly used to explain such a difference.

How can this biological ensemble of postulates now be incorporated into an interdisciplinary philosophy? First of all, one can consider these abilities as an expression of a special capability, one that is embedded in the continuum of evolution in the whole of life. One can also consider these abilities as an aggregate of such typical human characteristics as a *capability to distance*—an aggregate that neither describes nor postulates a "free will," nor does it depend on "free will" in an absolute way. Thus, the actualization of freedom provided by that capability to distance is an actualization of autonomy, but not of radical autarky. The difference is that autonomous processes, like organic self-regulation, only exist because of a vast range of ongoing and spontaneous interactions with their environment.[10] For me, the tendencies of flexibility and universality are autonomous processes par excellence: They may change and reset interdependencies but cannot (autarkically) live without them.

However, sooner or later, the whole of philosophy revolves around the question of freedom and, obviously, human capabilities have to be brought into play at that point. But it is remarkable that only a handful of philosophical discourses find in biology, history, or evolution essential aspects for understanding the concept of freedom, and human capabilities expressive of freedom, at all—and not merely as a use-case for ideas, theories, and systems. Counterbalancing this deficiency is one purpose and distinctive attribute of a bio-philosophy, and this attribute relates to a strong aversion to stasis and rigidity, which is not adequate for life and evolution. Seeing the fundamental necessities not in logical postulates or ahistorical categories, but in life itself, also means conceiving them in the light of their continuous changes. Solely due to this, freedom cannot be considered as an unconditional and immobile notion that operates its regulatory business independently of all bio-processual relations. A prominent advocate for this bio-philosophical mission is Henri Bergson. This article merely aims to consider his teachings on the human capabilities of instinct, intellect, and intuition found in *Creative Evolution*, although other aspects of his philosophy are relevant for the question of humanness as well.

Excursus: Bergson's Teaching of Capabilities as an Answer to Whitehead's Claim Towards a Speculative Philosophy

There are, among others, three general reasons why even today Bergson's teaching holds auspicious answers to Whiteheads interdisciplinary claim of a philosophy as a flexible and coherent system of general ideas. Thus, they are not only reasons, but real desiderata of philosophic inquiry:

1. **Dynamic Capabilities**: A first reason appears if we look at Suddendorf's conclusion. He carefully states that humans are equipped with *open-ended* abilities of reflection and mental representation. What a brilliant choice of words! He does not write about sheer *infinite* and *unlimited* abilities, although, unfortunately, some poor official translations have done so (e.g., the German version of his book). I believe that *"open-ended"* is semantically nuancing an unpredictable and unprecedented development as well as usage of these abilities and their results. Thus, our mental equipment is unbiased as to the result, and therefore full of *creative* potential to bring new and unknown occurrences into the world— well, at least into our imagination. The words *"unlimited"* or *"infinite"* seem to be inappropriate, not just for Kantians, to describe our mental abilities. *"Unlimited"*? Somehow that would sound like a rhapsodic characterization of the gap between humans and other animals, which, in consequence, seem to have only limited abilities. But, of course, this is only one side of the story. Although the so-called gap seems to be an absolute dichotomy, we can't ignore the evolutionary biological approach that human-animal differences are less fixed and solid, but fluent—not only for the body, but for the mind as well. In other words, every postulated

limitlessness must not end up in an inadequate glorification or a false dream of humanness. Therefore, a recourse to Bergson's *dynamic* teaching of capabilities is valuable in that it can subsume the ensemble of abilities mentioned under the general terms of *instinct, intellect,* and *intuition.* His ideas enable us to emphasize the particularly human qualities without dispensing with the open flow and continuous processes of evolution.

2. **Capabilities as Tendencies**: Another aspect of this agility appears when Bergson labels his general terms for the capabilities of living organisms (instinct, intellect, intuitions) with the word *tendencies*. In my opinion, Bergson's concept of tendency is most valuable for process thinkers, but it often falls short of philosophical attention. Maybe it is less noticed due to other prominent, important, and process-teleological classics like *durée* or the *èlan vital*—as Alex Gomez-Marin points out in his inspiring chapter.[11] And yet, for me the value of the concept of tendency lies in its significant role as a mediator. Its task is one of the most desired by process philosophers because it is supposed to build bridges between apparently separated realms, and it allows for a blurring and suspension of disliked and static demarcations (that are often claimed by classic metaphysical approaches). In this case, it is about the demarcations between concepts and reality. Here is one example: to say that the concept of a capability like the intellect denotes several tendencies means, i.a., to say that the so-described entities of the intellect do—in fact—not necessarily have to occur in a pure way. To say that the intellect *tends towards* consciousness while the instinct *tends towards* unconsciousness does not mean that all factual intellectual performances which can be found in concrete animal and human interactions are purely conscious, while the instinct is purely acting unconsciously. According to Bergson it is even possible that this *pureness only exists in the conceptual form,* but is never part of actual living processes. It is possible that the conceptual discrimination of what is the pure instinct and the pure intellect does not meet the real world. To avoid metaphysical concepts, which could be dangerous and misguided abstractions, but not to give up the clarifying benefits of their generality, Bergson uses the word *"tendency"* to bring together the metaphysical pureness of instinct and intellect with the variety

of concrete and real intellectual-instinctive performances. In other words, in most intellectual performances a bit of instinct is mixed (and vice versa)! Thus, introducing capabilities as tendencies can not only secure the procedural and evolutionary description of its performances, but can also preserve its diversity, which in the end belongs to forms of biodiversity and merging biological processes. This strategy is important in avoiding a universalistic, static discursive *"order of things"* (thinking of Foucault) that due to its abstraction becomes an inadequate reference to the real-world processes and phenomena by stating false, empty, or gross conceptual distinctions.

3. **Capabilities and Social Criticism:** If we agree that human mental abilities are not unlimited, but extremely flexible, and that this is one of many conclusions and satisfying ends of a scientific enquiry into the human condition, according to my introduction, then we cannot be satisfied yet. To philosophers there must be more, because, after all, that human flexibility *can be dangerous,* too. It is our role to extend the descriptive business of scientific exploration of mental abilities by normative considerations, or, should I say, by *criticism*. Therefore, bio-philosophy is supposed to be more than a lame follower of sciences. Bergson's conceptual framework is not only useful for descriptions, but it also provides us with a framework for criticism: a broad range that belongs to an interdisciplinary claim of a philosophy as a flexible and coherent system of general ideas as well. *At least*, the human intellect is able to transform our own biosphere in a way which is hostile to life in general. With Bergson, it is possible to show how contemporary social criticism (and so a criticism of established power relations, or the human modes of actions that can be realized within these power relations) can begin where we speak of those mental capabilities that are responsible for the actualization of the dangerous and surely not always endlessly admirable abilities and actions. Bio-philosophy as a bridge between biology and social criticism, so to say.

Instinct and Intellect as Evolutionary Capabilities of Closeness and Distance towards Life

First, let me offer a brief characterization of instinctive and intellectual tendencies in the evolution of living organisms. According to Bergson, an organism which has instinctive and intelligent abilities generally uses them to interact with its environment. These interactions are mainly described as *'knowledge,'* creation, and usage of tools.[12] In my own words, the instinctive usage of tools relates to a *tendency to increase closeness* while the intelligent usage of tools relates to a *tendency to increase distances.* Bergson characterizes the instinctive exposure to tools as *incarnate, facile, specialized,* and *latent* with a tendency towards a reduced consciousness. It is incarnate because the instinct has an *organic* closeness to its tool—here, we can think of an organism, which, in certain situations, uses its own organs according to the functions and abilities it has. It is facile in the sense that the instinctive operations are exerted with promptness. It is specialized because instinctive performances tend to align with concrete objects. And finally, it shows a tendency towards a reduced consciousness in the sense that there is only a small distance between cause and action. That means that the instinctive *"knowledge,"* by reasons of this smaller distance, is *latent*. On the other hand, intelligent exposure to tools is characterized by being *less incarnate*, straining, universal, and *tending towards an increased consciousness and an inorganic mode of operation*. It is less incarnate because the intellect has abilities to build and use distal tools that are not only organic (i.e., elephants using branches to acquire food). Though there is another form of distance implied: the intelligent tool is not only distal but can also be built indirectly by other tools. The intellect enables the fabrication of tools to build other tools to build tools and so on. Thus, the intellectual exposure of tools is not as facile as instinctive operations, but straining. The expansion within the fabrication of intelligent tools means higher forms of investment. A good example of a distance-related investment is the human way of reflecting before acting. It seems to be magnificent (but not always wise) how much distance we can bring between possible activities and the realization of one of them! The intellect is not only a capability to bring a certain distance between intention and action, this distance also constitutes an active opening of a space of *possible* actions. While the instinct tends towards the necessity to act according to an immediate life-serving or practical use, the intellect, due to its formal and from materiality abstracted, knowledge and tools is able to achieve the freedom to be flexibly prepared for several concrete situations that *can* (but do not have to) occur in the near or remote future. Hence, this example also shows intelligent performance as tending towards consciousness. It requires more than *latent knowledge*: what is needed is a certain internalization of the possibilities of actions. Furthermore, the intellect is not aligned with *concrete* objects, but with universal *relations* that enable a performance in multiple possible, concrete cases (think of the famous hand ax whose intelligent *form* offers a versatile usage). Furthermore, this flexibility can be exemplified by the most universal, powerful, intelligent human tools: symbols and concepts! Referring to Jakob von Uexküll's concept of a *functional circle* (*Funktionskreis*), Cassirer already saw human usage of symbols not as a tendential, but as a schematic performance to increase distances to the organic world. Thus, symbols are a third link or a typical human network (symbolic system) that takes a place between the usually and animally interwoven organic receptor system (which receives outward stimuli) and the organic effector system (which reacts and responds to the stimuli). But if a symbolic system is in charge, an entire life-changing *qualitative* difference occurs between organic responses and human responses. According to Cassirer, this difference marks a human dimension of reality, namely, our symbolic universe. Through the symbolic distancing in human activities *"the answer [response to stimuli] is delayed. It is interrupted and retarded by a slow and complicated process of thought."*[13] Conceptual thinking is thinking *about* something. The term *about* is a proxy that shows the objectivating, relativizing, distancing within human interactions with the world, others, and itself (for example, by conceptually referencing to its own feelings). Surely, an intelligent capability to distance by using inorganic tools can be found in animal intellects as well, but in the case of humans it operates in a special, symbolic way. Therefore, human is not only the *animal rationale* (intellectual tendencies are manifold in the animal realm), but the *animal symbolicum*. Cassirer evaluates this symbolic human discontinuity mainly as a scheme, but to bring it back to the processes of biologic evolution, one should consider the work of anthropologists like Terrence W. Deacon.[14]

The intelligent-symbolic discrimination is especially useful for us to get along with the world and to interact with our environment. It is important for all the earlier mentioned abilities to bring distance between the motivation to act and the actual action, to reflect *upon* something, to choose the best solution strategy, to set and weigh purposes, to plan and think ahead, to act sustainably,[15] to tell a joke, to use irony, to laugh *about or with* someone,[16] to ask a question, to problematize something, to know something or to be able to generate propositional and theoretical knowledge about the world. According to Bergson, the ability of *speculation* is the highest form of distance from organic life processes, because it can even abandon actual needs and necessities for immediate actions. Fictive scenarios, meditations, possible worlds (and even brains in tanks), utopias, and dystopias do not necessarily involve immediate actions. Perhaps, but only perhaps, we will need them or will derive benefits for our actions. Speculations can provide us—in the present—with possible actions for future scenarios; they can equip us for the future. Maybe they are meant for higher purposes, like the transfer and criticism of cultural traditions and mindsets and, of course, the intersubjective exchange of individual or collective views of the future. Fictive scenarios especially extend our presence to the future: an active distance to the present in the present becomes one strategic way of setting purposes and paths to shape the development of ongoing processes. In conclusion, intelligent distancing is one way of giving sense to the future. It is one way to establish a meaningful relation between the past, how it is incorporating and anticipating in the present, and the future, how it is streaming out of the present. But that is only one side of the story, and it can become dangerous if it appears to be the only way, as I will try to show now.

If the Intellect Exceeds Itself: Danger to Life due to Distance to Life

Dear human, consider yourself using your intellect to invent an inorganic tool to increase your mobility; let's say, by a technology such as a car. Mobility, indeed, is a natural, basic need of many life forms, of yours too. Granted, this motorized, artificial enhancement is not only a tool to increase your mobility, it also became a *symbol* of your freedom. It is a masterpiece of intelligent performance, but it is unbelievably stupid, too. During some days, due to traffic's CO_2 emission, the air pollution in Beijing reaches such a level dangerous to human health that the government advises you to stay at home and to close every window. What began as a technique to increase your mobility turned into a tool that is now somehow massively restricting your freedom and possible movements. The capability to create and use intelligent tools does not imply that the realization or usage of these tools always is clever. Although their evolutionary purpose in general is life-serving, these tools can become life-threatening, especially if their universality and flexibility are distancing too much from their evolutionary connection and so from life. According to Bergson, the intellect tends to distance itself from life because it is enabled to apprehend and manipulate the world with its own *mechanistic* (non-organic) mode of operation. Furthermore, intelligent technological tools and artefacts tend to manipulate the world (machines as well as symbols, norms, etc.), i.e., by establishing dependencies and power relations to canalize possible actions in the social world or to canalize the ways in which we generate our knowledge in daily life and scientific discourses. These intelligent performances can not only restrict human freedom in various life-unworthy ways, but can transform our environment into a realm hostile to life. This transformation of intelligent tools now hostile to life can be seen as a *pathological actualization of the capability to distance*. The intellect has lost its contact to life in a dangerous way. Although he never jumps out, Bergson opened a window to an extensive criticism of intellect and its social-ecological impact on well-being. He is in good company: Adorno's and Horkheimer's *Dialectic of Enlightenment* criticizes the totalitarian and (mass-) unitizing essence of human reason, Husserl's scientific and philosophic European crisis criticizes the lifeworld loss of meaningful subjective experiencing due to the primacy of mathematical and formal abstractions or a technologized world. Bergson's earlier conclusion simply is more general in saying that *the intellect tends to pretend to be the only way to apprehend the world*. The primacy of intellect is more than only a primacy of reason and rational principles, because it is connected with an organism's possible loss of life experience. It is the idea that an intelligent oblivion of instinctive and sympathetic closeness to life is the crucial moment where processes of future human development can get an unhealthy and most dangerous drift, and, far from life, they can become senseless by losing their purposefulness.

Closeness Through Distance: Intuition and Creativity as Life-serving Correctors of the Pathologic Intellect?

It is possible that Bergson guides us to his own solution of how the dangers of intelligent tools can be contained. What is it that can be done against an inadequate, out-balanced intelligent distance? Well, a bit of closeness could be helpful! For that, the intellect can fully regard and benefit from its own abilities. Thus, Bergson says that the intellect is even able to *distance itself from itself!* Since the intellect recognizes its tendency to the mechanization of and distancing from life, it asks other capabilities for help. Thus, the organic instinct is much closer to life than the intellect. *The capability that can join both the intellectual and the instinctive operations could be called intuition.* For Bergson, intuition is the intellect's attempt to prevent its own tendencies from exceeding themselves by searching for an extension of instinctive knowledge much closer to the experience of life. Via intuition, the intellect mobilizes the instinct to search for the lost closeness to life. Due to its inorganic mode of operation, the intellect is not able to *know* much about organic processes, but thanks to intuition it can grasp a *hunch* of it, handed in by instinct. To be precise, Bergson's intuition is not a separate, third capability, but more an in-between mediator. It can only do its work with the tendential unconsciousness of instinctive operations and with the spontaneous operations of the intellect, whereby the intellect is the driving force for every intuitive endeavor. Why, after all, should the instinct search for closeness to life? It *has* closeness and sympathy with life; furthermore, it lacks a spontaneity to initiate such an endeavor. Intuition is apprehending the world somewhere between feeling and thinking its contemporary character. But what is in charge is not the world's character as a static being (its actual state which is abstracted from the lively movements that factually happen without pressing a pause button!). This abstraction can be done by the intellect alone. It can, for example, subsume and conserve states of the world *as inert matter* via concepts (that is, how we *analyze* and *think* the world). The intuition is about understanding the world in its current processes, in its becoming: a natural, organic understanding of life which can be more than just instinctively *felt*, but *experienced*, in terms of *it can be lived*.[17] Here is the crucial paragraph:

> Instinct is sympathy. If this sympathy could extend its object and also reflect upon itself, it would give us the key to vital operations— just as intelligence, developed and disciplined, guides us into matter. For [. . .] intelligence and instinct are turned in opposite directions, the former towards inert matter, the latter towards life. Intelligence, by means of science, which is its work, will deliver up to us more and more completely the secret of physical operations; of life it brings us, and moreover only claims to bring us, a translation in terms of inertia. It goes all round life, taking from outside the greatest possible number of views of it, drawing it into itself instead of entering into it. But it is to the very inwardness of life that intuition leads us – by intuition I mean instinct that has become disinterested, self-conscious, capable of reflecting upon its object and of enlarging it indefinitely.[18]

Bergson's metaphysical ideas about intuition are extraordinarily prospective, if one considers the state of scientific research at the time, which was, after all, a hundred years ago. Bergson's intuition is mediating a shift between different *qualities* and tendencies of closeness and distance, i.e., different qualities of unconsciousness and consciousness, while scientists obviously tend to transform those qualities into quantitative differences (scale from zero conscious to max conscious, etc.). Nevertheless, one can find astonishing similarities in the vocabulary. The psychologist Scott Barry Kaufman states that intuition is operation in a very special mode: *"Intuition arises from unconscious, or spontaneous, information-processing systems."* This interesting synergy between both systems is a *"delicate dance of analytical and unconscious processing."* Unconsciously motivated, we intuitively *"tap into the conscious, rational mind to play around with the ideas we've generated, and to uncover their uses."* Not only the mode operation, but also the meaning reminds us of what Bergson has already said. Thus, Kaufman continues: *"Most problems in Life require insightful thought processes."* By being close to life, intuitive abilities are most important for the assimilation of new information into our existing memories, our existing knowledge, etc. They enable us to perform faster than just intelligent responses in confrontations with new problems and problematic situations by evoking bygone experiences in which we used *analogous* or *similar* solution strategies. On the one hand, the intuition extends far into the performances of the intellect: intuitive insights can become unexpected shifts (far beyond cognitively controlled derivations) in our processes of understanding. On the other hand, Kaufman emphasizes the problematic, but relevant emotional aspects while having intuitions. Positive and negative emotions can become a highly motivational intensity before, during, and after intuitive moments—an intensity that one probably will not find in analytical problem solving.[19] It is remarkable how Kaufman reflects this *"delicate dance"* of intuition as being a *strategic* interplay between two *antagonistic* and *unlike* forces, which reminds us of great process and *différence* thinkers like Nietzsche, Foucault, Deleuze, and probably Bergson himself.

What this all amounts to is that even in contemporary scientific inquiries, one idea remains immanent: via intuition, as one part of human creativity, our intelligent abilities maintain connected with life, its subjective experiencing, and the closeness, immediacy, or life-sympathy of the instinct. Surely, therefore, intuition and creativity can become and always were a correction of the possible failures of our intellect. The intellect's own tendency to be the only adequate or primary way of interacting with and objectively knowing the world is a natural *tightening* of the instinctive and intuitive diversity of worldviews and experiences. *For the intellect, tendentially subjective access to the world primarily appear as opponents.* But, then, intuitive moments remind us of seeing this diversity not as an opponent, but as a supplement.[20] And perhaps it is even more. Speaking with Husserl, it might be the case that intellectual objectivity cannot create *any sense* without arising from a subjective and lifeworld experience. This is how Bergson's intuitive interplay between instinct and intellect, between feeling and thinking, becomes a most important feature for meaningful processes which prevent our intelligent performances from turning into pathological, dangerous, and *senseless* (also stupid) strengths against the world, other living beings, and of course against ourselves. Thus, Bergson's intuition should be considered as a capability that is interesting to social criticism. Eventually, it turns out to be a key competence to correct human courses of action. *"Acting reasonably"* does not exclude our emotions, and it does not exclude intuition as merely a vague premonition—not if that apprehension is able to bring back closeness and sympathy to life that the intellect tendentially does not seek. Is intuition a *typical human* corrector of the intellect? Maybe the empirical and scientific answer must wait until it can be excluded that these interactions between automatic and controlled mental processes do not occur in animal life. Though, in Bergson's conceptional perspective and due to *the necessity of an intellect enabled to reflect upon itself to reach life again*, intuition seems to be a quite human condition, perhaps one that will be more developed in a *future* humankind. After all, humans are the only living beings on planet Earth whose intelligent abilities impend so dangerously distancing to life. The closeness that intuition can offer does not have the same quality as the closeness of instinct. It is a different tendency to closeness because in intuitive insights the closeness to life is *likewise mediated* intellectually (intuition as self-conscious instinct). Contrary to instinct, intuitive closeness tends towards consciousness. Even if it sounds delicate, I desire to say that the newly won closeness to life in intuitive moments *emerges* from our intellectual capability to distance: intuition as a closeness, mediated by distance.

HUMANS AS PROBLEMATIZING ANIMALS

What has all of this to do with power or Michel Foucault's power concept? At the beginning, our idea was that the just evaluated human capabilities shape our spontaneous actions and canalize the possibilities for action. Normally, these actions take place in social spaces and therefore within mutually influencing dependencies and power relations. As Spyridon A.

Koutroufinis points out in his chapter, the design and interaction with our *Umwelt* is bound to the experience and contact with symbolic and normative relationships. Furthermore, our symbolic and normative relationships are traversed by different modes and occurrences of power exertion. As mentioned earlier our symbols are, in fact, *powerful* tools. Thus, when talking about a typical human *Umwelt*, we should include the human experience of power and how we design mutual dependencies for our interactions. A very volatile thought is that the most effective powers rely on a tendency to avoid being consciously experienced by affected subjects. We are now confronted with the problematic relationship between our freedom of (spontaneous) action and power relations that seem to act on that freedom. If we are permanently surrounded by and embedded in power relations, they must have a heavy impact on our intelligent and intuitive management of modes of thinking and acting. I know most of you will share the intuition that evaluating our freedom of action in the light of power is a devious enterprise. Often freedom and power are recognized as antonyms: Where there is power, there is no freedom; where power is enlarging, freedom vanishes. Nevertheless, Foucault's unique concept of power can open up an opportunity in which *freedom of action becomes a constituent moment of power exertion*. This power-sized freedom of action complements the freedom in our mental capabilities to distance.

Towards a Friendlier Face of Power

First, I would like to give a brief summary of the most important generic features of the power concept. Foucault once summarized his concept of power with the words: *"Power is not an evil. Power is strategic games."* [21] Power is not just a notion of the totality of an all-repressive mechanism, nor is it some-*thing* that can be possessed by persons or institutions, something they can use to exert restraint on other persons. In fact, all these repressive mechanisms are important. But Foucault also insists on a friendlier face of power: its productive and creative potential. Furthermore, power is always an in-between, so it needs to be between at least two individuals and not just held by one. That is why power is always a *relation* in the need of at least two sides: the act of exertion itself needs playmakers as much as takers. Later, this characteristic will reveal an interesting interpretation and the necessity of practical freedom, especially on the side of the taker! The culmination of the idea that power relations have repressive-productive effects can be found in the first volume of *The History of Sexuality*.[22]

Power names a diversity of strategic situations, games, and flexible, agile relationships within societies. Thus, social dynamics can be analyzed:

1. as strategic games, fields, and situations;
2. as a variety of relations of strengths: strengths with repressive-productive, destructive, creative, organizing, and transforming effects, and mechanisms that influence actions, reactions, and settlements of different purposes;
3. as flexible relations and ratio of interdependencies; as mutual penetration of heterogeneous, local, and instable games of regularities;
4. as simultaneously intentional and non-subjective movements which can be further characterized by the analysis of resistances, discontinuities, and bursts or strategically designed transformations.

The features of power seem to be quite flexible. Words like *"instable," "local," "transformative," "creative," "destructive,"* and *"mutual penetration"* show power as a very processual concept. In that sense, it has much in common with the concept of *life*. Provocatively speaking, the concept of power is a concept of living. In his article *The Subject and Power* (1982), Foucault implicitly names two necessary conditions for the possibility of power relations: *sociality* and a special kind of *"practical freedom."* [23] First, power relations can only be established when there is a kind of socialization, which is to say, there is a need for another individual. Second, power can only be exercised *by* and *over* individuals who possess a certain amount of practical freedom, so that the individual possesses several (more than one) options to behave, react, or act. There always must be spontaneity on *both* sides of the relation. Otherwise, a lack of such freedom or options would impede the establishment of a power relation in support of the exertion of *violence* (so power is not equal to violence, but violence is the most radical and repressive mode of power operations). Foucault concludes that every form of exercise of power can be analyzed as an action and so be described by a vocabulary of intentionality, the settlement of *purposes, choices* or acts of *choosing between alternatives,* etc. To be more precise, *power* is exactly defined *as an action upon an action of another individual.* This relatively open space of possibilities that is necessary for the unfolding of power doesn't *necessarily* involve questions about the individual's consciousness. It simply does not matter primarily if the individuals whose actions are canalized by the power relations *know* or *reflect upon*, *desire* or *misprize* those influences on their actions. Power has a *seducing* effect on possible actions. It *creates* appeals and needs, and it narrows and widens possible actions by being performative and procedural. Power affects possible actions by operating on the performance of choice. In this sense, practical freedom is a collective term for heterogeneous, manifold *dispositifs* of regulated expansions and constrictions of possibilities within a game of interdependencies. It is important that the leading sign of power towards freedom is not always a negative sign, but also positive according to the creative potential of power exertion. Especially *the feature of choice and alternatives* as a necessary condition of the power mechanism to operate most effectively (which is not the case in violent and most static power exercises) requires subjects which might be manipulated or even seduced and canalized in their actions, but never powerless, and so never acquitted of acting with a purposeful self, using its intelligent and intuitive abilities.

Power and the Self

If we draw a line between Foucault's and Bergson's basic concepts, it can be said that the establishment of power relations mostly rests upon intellectual performances. First, the so important notion of strategy in power arrangements could be understood as an extension of evolutionary intellectual tendencies towards a conscious purposiveness (a tendency towards purposive activity). Still, we should remember that power with such a tendency does not always have to be exerted consciously. Second, especially human power relations are often arranged by symbols, but also exerted with an intervention of other, for example, technological, inorganic tools. Thus, power relations are performances of distancing. Third, the distance from immediacy is necessary for the freedom of action within an alter's spaces of action. For Foucault, this freedom is a condition of the ego's possibility to repressively and productively influence the alter's choice and actualization of interactions with the world and with others. Both the alter's and the ego's abilities of universal, flexible reflection and the determination of aims are important to exercise power. Although we might say that instinct, intuitions, feelings, and desires are important sources of the ego's wish to continue its own purposes in alters' actions, it often is the intellect that becomes the strategic instrument to manage these concerns and objectivate them, release them from pure subjectivity, and enforce them to take a place in intersubjective, social spaces. Somehow the intellect is a capability to realize human freedom, but of course it can therefore be an instrument of power, influence, and domination (as is reason). It is a capability to regulate mutual canalizations of possible actions; it organizes the sheer diversity of human desires and purposes. And if you think of Foucault's famous modes of disciplinary power, these orders really can be a mechanization and rational shortage of plurality and diversity of life and living beings. Think of school systems, health care, and military barracks whose disciplinary orders mean an enforced conformity.[24]

Often, but not always, human subjects do respond *critically* to these mutually influencing processes. They start to explore these dynamic power games and position themselves against them, often with the aim to transform their own dependencies. Most modes of power exertion try to prevent their exploration as well as to discourage the concerned subjects from positioning themselves against them. This can happen in various ways. The most radical options are staying invisible (avoid resistance), or violently enforcing the alter's powerlessness by minimizing practical freedom (power turns into violence and so violates its own condition of possibility: the alter's practical freedom). As a result, human subjects often either have no interest in exploring their dependencies, or they do not have a chance to do so. If we now turn back to the question 'What is the human?', we simply cannot add that it is an animal who explores its own power relations. Since this fact *can be vanished by power*, a critical idea of the human condition must turn this fact into an imperative of humans. It is a task we must continue to undertake! The positioning against and the transformation of persisting power relations requires a subject that constantly redefines its own self and its autonomy. Thus, the self as well as its autonomy is not a human fact that can be stated, but a human process we should steadily *take care of*. The later works of Foucault focus on this care of the self. Although he refers back to ancient times, he worries about our present: maybe we are embedded in power relations that dangerously keep us away from developing a critical self.

The Care of the Self as a Practice to Fulfill Freedom

The processes of becoming a subject are always embedded in persistent power relations even though there is an indeterminate nature of the self, enabling spontaneous feedbacks. In his later works, Foucault investigates the so-called *'care of the self'* which forms an ensemble of self-practices in the Roman-Hellenistic society of the first and second centuries (AD).

For him this era is a climax in the care of the self because it is integrated into an art of living that involves the entire margin and every domain in the life of nearly every individual within the society. The most important function of the care of the self is the development or improvement of one's own ability to set purposes. The goal is nothing less than personal salvation. It is not just a matter of escaping from certain dangers or threats, or *"escaping from the prison of the body, escaping the impurity of the world"*; it is also a matter of enabling oneself to be prepared, to be in a state of resistance, a state that enables and saves one's sovereignty. It is salvation as the state in which you do not have to fear the danger of being enslaved or dominated. It is a state in which you yourself will be responsible for your own happiness. So, again, the aim of salvation is not solely a spiritual attitude, but it requires an ensemble of practices of the self—even medical care is involved. Often, salvation is the aim of relieving or freeing oneself, of honoring and respecting oneself.[25] Becoming a subject is not a matter of becoming someone out of nothingness, nor is it a transformation that follows one final course or reaches a final end. Becoming a subject is bound to multiple creative and destructive changes or modifications. The unfolding of the subject does not, as power relations do, merely point in one direction. There is a constant reorientation within the procedural dynamics that are driven by the practices of the self. Even salvation as a general aim of those processes is changeable and can anticipate different forms or demands. What are ancient moments of becoming a subject? There is a fascinating and, at first sight, cryptic comment on Seneca that is a perfect hodgepodge of the vital elements of these processes of the subject:

> Inasmuch as stultitia is defined by this nonrelationship to the self, the individual cannot escape from it by himself. The constitution of the self as the object capable of orientating the will, of appearing as the will's free, absolute, and permanent object and end, can only be accomplished through the intermediary of someone else. Between the stultus individual and the sapiens individual, the other is necessary.[26]

Hence, the transfer from individual to subject refers to a transformation of the nature of the mind from a determined to an undetermined settlement of purposes. The determined mind is characterized as a *"nonrelationship"* the *"stultus individual"* has to its own self. Thus, the mind of the *stultus individual* is distinguished by restlessness and irresolution, which Foucault calls the raw material with which self-practices have to deal. It is important to note that this *stultus* mind is generously open to external influences or their unaudited and unfiltered intermixture with its own wishes and ambitions. So there is a lack of a *discriminating* apprehension and prehension. The inability of discrimination or, if I may, the absence of a certain type of capability to distance, impedes a necessary concentration for a will that can set its own purposes. An untrained will is always determined by external influences, so it is not free and cannot want absolutely or permanently (so it is not possible to follow one purpose for a longer period). But there is more: the *stultus* is not just unwillingly open to the world, the will is also lost, dispersed, or broken up in time: *"The stultus is someone who remembers nothing, who lets his life pass by, who does not try to restore unity to his life by recalling what is worth memorizing."*[27] It is interesting how similarities between Seneca and Bergson occur in those few lines. Both bring together an unfree will with an absence of discriminating abilities, which is analogous to a form of capability to distance. I just recall Bergson's intelligent tools of human concepts and symbols, which enable a discriminative comprehension of the world by being most universal—a universality to enable an optimal interaction with and within the environment. this fickle mind will be overcome only if an individual trains a specific relationship to itself. Only then can the *Homo stultus* become *Homo sapiens,* with an organizing, relatively steady, and orientated will. To provoke again by using Seneca, the difference between being human and being an animal is the difference that I called a specialty in the human capability to distance, which includes the possibility of becoming a subject. This difference is not simply an unmovable human *fact*, but it is, indeed, a human, realizable *task!* The quotation names another interesting element of this task; namely, the necessity of someone else. In the end, that strange quotation reveals four highly interwoven moments within the processes of becoming a subject:

1. The ensemble of self-practices contains not only practices which *constitute* the subject or *setting oneself as a subject*...

2. ... but also practices to *correct* the subject by experiencing its environment.

3. Those practices directly motivate the movement of several processes which include the role and influence of other members of the society...

4. ... but also a conversion to the self. The *"care of the self"* is a combination of a specific mode of *ethopoetic knowledge* (*mathesis*) and a specific *ensemble of practices* which can be called *askesis*.

The first moment is the general form of the process of becoming a subject, which was described as the transformation from a *stultus individual* towards a *sapiens subject*. The second moment declares the heterogeneous character of the transformation. Becoming a subject never ends but always needs critical re-examination to correct mistakes, errors, bad habits, and misleading and undesired interdependencies that have possibly been established and are now engrained.[28] To guarantee a successful transformation the influence of an other is inevitable. Within the Hellenistic-Roman epoch, the other, of course, is the philosopher who appears as teacher or as personal counselor. But more important is the social factor within this necessity: becoming a subject always requires a special mode of conversation which Foucault names as *parrhesia*, translated as a form of *liberated* speech. The fourth moment is the conversion to oneself. To turn to oneself does not mean to entirely turn away from the world or from other members of the society; it is not a contemplative, egoistical world withdrawal. On the contrary, the conversion comprises both poles as core elements. In fact, the conversion is a matter of positioning oneself towards a better relationship between the self, the world, and others. An adequate relation to oneself *is* an adequate relation to the environment. The conversion contains a turning away only from certain aspects of the world, but under which criteria? Again, distance and an ability to discriminate is required, but there is more. Obviously, this distance must become closeness; it must bring back the objectivated knowledge to the living, experiencing, and meaningful subject. On this occasion, discrimination is characterized by a certain mode of knowledge, a *mathesis*, and a specific mode of actions or practices, *askesis*. In general, this mode of knowledge always combines the truths which it contains with prescriptions towards someone's behavior. It is a knowledge of *"prescriptive facts."* It is a knowledge with no distinction between the truth and the subject's relation to it; a knowledge about how these truths correlate with the practices of self-regulation. Therefore, it is a knowledge about how *logos* and *ethos* cohere and so can be called *"ethopoetic knowledge."* Knowledge becomes a calculation of usefulness and a calculation of liberty. How so? Demetrius gives an example; the conversion of self which enables and requires ethopoetic knowledge is a distanciation to the world. It is an act of discriminating useful knowledge about the world from pointless knowledge.[29] Follwoing this idea, it seems pointless to know that gluons and not quarks are causative for the spin of protons, while it is useful to know that you live in a society whose social system will support you in awkward times (at least in some countries). Not *causal* knowledge but *relational* knowledge matters. The *conversion* requires training a capability to bring distance between the knowledge about the world and the concurrent valuation of this knowledge according to oneself. Furthermore, in Epicurus's texts, this discrimination of useful, prescriptive facts is fundamental to changing your mode of being in the world and thus to human liberty. It makes you self-confident, happier, brave, and resistant. Maybe this operation of care can be described as re-embedding objective knowledge into subjectivity, into meaningful, living experience. To accomplish this immediate transfer from logos to ethos, to transfer facts into prescriptions, an ensemble of practices (*askesis*), which Foucault subsumes under the concept of *paraskeue*, must be internalized. Thus, practices of listening exercises, exercises of reading and writing, speaking, and the examination of conscience prepare and equip the self for possible happenings in the future.

"There is no First or Final Point of Resistance Other than in the Relationship One Has to Oneself"

All in all, the processes of becoming a subject means examining and evaluating one's the entire life. Experiencing life, experiencing the world, the environment, the bios, must correlate with a specific kind of techne. Implicitly, Foucault is fascinated by these ancient practices and criticizes the way in which they have become forgotten and what such a loss could mean:

> even though it may be an urgent, fundamental, and politically indispensable task, if it is true after all that there is no first or final point of resistance to political power other than in the relationship one has to oneself.

He continues:

> I do not think that reflection on this notion of governmentality can avoid passing through, theoretically and practically, the element of a subject defined by the relationship of self to self. Although the theory of political power as an institution usually refers to a juridical conception of the subject of right, it seems to me that the analysis of governmentality—that is to say, of power as a set

of reversible relationships—must refer to an ethics of the subject defined by the relationship of self to self.[30]

And here it is, the fundamental passage where we can see a human feature. What evolution of intelligence creates as certain modes of being, capable of increasing distances and so capable of living in power relationships, is now shown in a typically human guise. A human cannot only relate itself to diverse interdependencies, but the human can do it in a special, ethical, and critical way. Its reflection of interdependencies *is* and *should* (!) be based on a *"relationship of self to self."*

Building on this idea, I conclude this chapter with an ethopoetic knowledge, that is, a prescriptive fact. It is a fact that challenges us to self-limit our intelligent power performances; for example, by intuitively recalling their meaning for life and subjectivity. The fact of this human discontinuity is inseparably interwoven with an imperative or the political assignment (in its broader sense) to train the self in order to be able to exert influence on established power relationships. Thus, the human feature can only be *understood* ethopoetically. Alfred North Whitehead once said: *"To be human requires the study of structure. To be animal merely requires its enjoyment."*[31] This assumption is valid for power structures as well. Human states of being interwoven with power relations have enormous influence on our way of being in the world. But because we are capable of reflecting (studying, problematizing) them, every single one of us can become a source of creative feedback. This interplay is the essence of human social living under the light of repressive and productive power dynamics. It is one way to describe the reason why we are human and not like dancer. Likewise, it is a quite optimistic thought if you problematize established ecological, economic, social, and metaphysical relationships that are dangerous for any future living on planet Earth.

The analysis of ancient self-practices in *The Hermeneutics of the Subject* helps to provide a connection between a bio-philosophical determination of human mental capabilities and the problematic nature of power. Not only is the self the core of the processes of becoming a subject and the subject's center of spontaneity (for actions, imaginations, thoughts, etc.), but it is also a creative *oeuvre*. It is a work and the constitution of a self in the framework of critical positioning and influencing existent and future dependencies. Earlier we discovered our intuition as a capability mediating between intellectual distance and instinctive closeness to life. It is a part of our creative potential to correct a misguided intellect that has lost too much contact with life. *The care of the self is not only distancing reflection, but it is self-referential reflection and action to bring the subject intelligent performances closer to its subjective experiences.* The involved participation of others and power relations does not only tend towards objective results, but subjective or subject-constituting effects. Intuition can help to bring the world and others closer to the self. We therefore must say that to find those new ways of sympathy is a creative process. It brings new connections and transforms our relationships: between us, us and the world, and, of course, we change our own self, too. Understanding life as an examination does include the acceptance of existing power relations as the framework for our experiences and the experience of our self. This means mobilizing others, the world, and social dependencies to explore this self. The care of the self is not just a speculation; it is a care that must be lived in the subject's actions. To say, then, that the human is the ethopoetic animal is indeed a critical process and not an everlasting statement. The *conditio humana* is a task, and its crucial maxim is to *preserve human agility within its own power relations*. Thus, our main critique against existing power relations should focus on the question whether a specific mode of power is paralyzing or supporting our ethopoetic character and our intuitive as well as our intellectual capabilities. Once I called the misguided use of the intellect a pathological use. Maybe I should go on and call the resulting power relations pathological, as well, if they are constraining our ethopoetic processes in an inhuman way. Ethopoetic maturity is not supported by powers that operate and extend by recklessly extinguishing alternatives and possibilities for a transformation or redirection of processes, or by exclusively using aggressive and violent means and instruments. Such an incapacitation (!) should not happen in scientific, technical, political, ecological, or, of course, economic realms. Perhaps we can easily see many corrupt or pathological power exertions that actively prevent us from caring about our self. Especially in Western societies, extensive carelessness is supported by different factors. For some, it is in the postmodern hectic way of thinking, or an unfettered efficiency-oriented thinking. One does not have time to care about oneself in order to build a critical and resistant self, but time for a short, superficial fitness program, yoga in the evening, *"medification,"* and optimization of daily life performances. Maybe these are new forms of disciplinary devices that try to make us more productive and better slaves to capitalism and its economy. And this indeed would be a very old warning of social criticism. For others, these factors are given by a complexity of global economic and political systems.

Even subjects who are willing to participate in governance have to cope with an *incredible problem of clarity*. Another group of careless subjects arises within the technological transformation of our *Umwelt* and world. For many of us, the technological promise of salvation is heavier than the salvation of becoming a subject. The care of the self becomes dispensable due to the technological automatization of our lives. The human self in lost motion? In many future scenarios given by Linda Groff, just as in the Buddhist-Transhumanist-Dialogue that Sean MacCracken evaluates in this book, these ideas are of great significance. Some might even think that the problems of climate change and the ecological destruction of our biosphere can be technically solved by leaving the planet and colonizing Mars. We should really ask ourselves if this deserves to be called a *"solution."* At least it does not have anything to do with an ethical project. Besides, it evokes ethical problems. Who decides who will be left behind? Or, Mars, our next planet to ruin? It's an escape, not a solution. It's not just an escape from what we may ruin, it's an escape from taking care of our self. Science, and maybe some rich technocrats from Silicon Valley, can do that for you. They can build technology that is not for your purposes, *the ones you can set as an autonomous subject*. Some technologies can or already have become ends in themselves. For these people, the term *"intuitive technology"* refers to a *foolproof handling*. Instead, and after all I have said about intuition, shouldn't we understand *"intuitive technology"* as a technology that supports our creative, ethopoetic potential? *Sapere Aude 3.0!* Do you want to be a dumb or a *smart* user? Should helpful autonomous technologies ever antagonize our own human autonomy?

A modernized care of the self must deal with many new challenges. The subject does not only have to bring itself in a critical position towards the world, others, and itself, but maybe towards its own technologies, too. Furthermore, and thinking of sustainability, the subject's concerns must also integrate the well-being of future generations. Thus, the distance of the self to sciences, techniques, and politics should not become boundless, but bound to a *life-serving* and *lifeworld endowment with meaning*. Humanity's realization of freedom massively depends on our reason, but only if it does not lose an instinctive and intuitive closeness to life. That's why Whitehead calls reason not only a guarantor of freedom, but a promoter of the *art of living*.[32]

FOUR

Evolution and What the Intellect Makes of It

Alex Gomez-Marin

I should like to come back to a subject on which I have already spoken, the continuous creation of unforeseeable novelty which seems to be going on in the universe. As far as I am concerned, I feel I am experiencing it constantly.
-Henri Bergson (1947)

THE HUMAN BEING is a paradox.[1] We, a result of evolution, have developed the theory of evolution. Namely, the evolutionary process, in an unprecedented attempt, has been thought by one of its products—the bootstrapping is in place: the *explanandum* nominates itself as the *explanans*.[2] Yet, the concept of evolution is one thing, while evolution itself is another. Upfront, this chapter is an attempt to rescue Bergson's intuitions on *heterogeneous continuity*, his notion of *multiplicity*, so as to recover that which, being at the core of evolution, has been lost by our habitual ways of thinking about it.

EVOLUTION IN THEORY

To account for the production and common origin of diversity in life, two principles are put forth by the received view that dominates evolutionary scientific thought: variation and selection.[3] The first ought to be blind, the second retentive. Namely, new stuff is produced (without a plan), and then some of it is tamed and maintained. Under this lens, the evolutionary process is nothing but selected heritable variation. The universe becomes an engine for the production and conservation of novelty. It is a novelty that, in a profound way, is never really new. The unfolding of life, with all its complexity—believed to be a precipitate of simplicity—still astonishing, has now supposedly been rendered explainable and, one may even dare to say, explained.

The method of "trial and error" (initially an eighteenth-century technique to teach kids to solve algebraic equations heuristically[4]) has been erected—from tool to theory—as a fundamental proposition. Evolution is thought by means of the direct analogy with the centrifugal governor of the steam engine, or as bacterial navigation. Yet, framed with a static concept (as with any frame), it initially works as a useful scaffold soon to make its rigidity progressively felt, defeating its purpose. The model replaces reality—the living tree becomes dead wood. Life becomes some sort of dead matter whose organization defies explanation. Note how, when abstraction replaces fact, tentative description easily creeps into stubborn prescription.

The foundations of evolutionary biology (more precisely, natural selection) raise three conditions to be met: there must be variation, it must have adaptive consequences, and it must be hereditary, namely, capable of being transmitted so as to be maintained. That is, random mutations[5] occur, environmental contingencies select, and genetic material conserves. And so there is descent with modification — phenotype selected, genotype passed. The iteration loops over and over. Yet, selection *of* something is always selection *for* something. One can certainly ask how, but one must not forget to ask whither, nor whence.

Evolution is not Darwin's original idea. His is a proposal as to one of its essential mechanisms: that it operates by means of natural selection. We could get into a long, yet interesting digression as to what is really selected. Yet, I want to emphasize another equally important blind spot: evolution "by natural selection" *of variation*! In other words, selection does not explain variation, it utilizes it. In order for its force to be effective, natural selection draws its power from variation. The production of difference and diversity is required in the first place so that such differences may have a chance to be selected and then conserved. "The survival of the fittest" cannot take for granted "the arrival of the fittest." Paraphrasing the Catalan embryologist Pere Alberch: the winners, not the players, are decided by selection.

Variation—and with it, evolution—is postulated blind (the same for selection but, as noted above, that is not my focus here). Random or not, variation does not intend its effect. It is blind because it is unable to see what it is doing. And if it were able, it would not care. In other words, by means of "runs and tumbles," evolution finds a way without knowing the way. In sum, causes are blind, and effects are selected. In this way, this flavor of the theory of evolution seems to achieve its goal; namely, to find a common origin for the diversity in life, addressing the problem of the One and the Many, and without having to let in any of the vital spirits so successfully exorcised in the twentieth century.

These two cornerstones of the theory of evolution—variation and selection—represent an incommensurable effort of observation and synthesis, as well as an unparalleled feat of the human intellect at explaining how the diversity of plants and animals came about. It is a pity, though, that such effort missed the mark. The theory of evolution had a hidden and more important agenda, operating as an ideology (even as some sort of scientific pseudo-theology), the purpose of which was not only to eradicate purpose, but to leave no room for anything *creative* to enter the explanatory landscape. (And note that I am not thinking necessarily about the naive notion of an old bearded man creating *ex nihilo* from the heavens.)

Moving forth, it is indeed tempting to simply invoke, in conjunction with natural selection, the process of random mutations and then mentally run the algorithm to produce, with enough patience (and intellectual leaps bridged with imagination), heathers, and harriers, and horses, and humans. A fundamental limitation of such thinking is not so much the disdain for Wallace's suggestion to ponder the role of nature's appetition for an end. A less debated issue, and arguably more critical, is the process by which the virtually new becomes actual. If the core of evolution is "a difference that makes a difference and then stays," we must then inquire into the nature of such differences. How do novel organisms arise? What is really their *origin*? Ultimately: *what is the source of the new?*

Evolution, as it is commonly thought of,[6] is not truly creative. And it is not creative because, according to our patterns of thinking, it cannot be. The new is merely a reconfiguration of the old—the same old bits, just spatially rearranged in a different manner. But perception seems to tell us otherwise: nature is ever surpassing itself. Most likely that is what Whitehead tried to capture by his use of the word *concrescence*: a bootstrapping out of a movement of differentiation, not of construction.[7] In a way, we are emphasizing, as Tim Ingold would say, "Evolution in the Minor Key," the view which acknowledges and celebrates a constant excess of novelty.

Before we proceed by the tight alley traced by the linear strings of logic, we need to unfasten an important knot. Note that by enforcing that any biological process remains unguided, we have condemned nature to be uncreative. The blind watchmaker cannot know, nor would it care to know, what time it is (and even less what time *is!*[8]), and so we humans now stand in front of the spectacle of our own evolution like a puzzled ghost with a PhD. Nothing grounds us anymore, not from above and nor from below. *To keep God dead, we have also killed Nature.*

Variation may be blind, yet there is a blind spot in our thinking about evolution. Do the two principles of evolutionary theory exhaust the phenomenon of evolution itself? Do we know what life *is*? A great deal of effort in the field of evolutionary developmental biology has been devoted to question the classic frame established by the Modern Synthesis. Unsurprisingly, Henri Bergson was a century head of his time.[9]

Note that to get rid of what already exists seems easier than to give rise to what is not. Variation can be seen as a process of expansion, while selection as a process of compression (recall the two "vitalistic" forces mentioned in endnote 3). Natural selection can be interpreted as more than a mere filter yet, in the strong sense of the word and in the mainstream view of its role, variation is not, cannot, and should not be creative. We are then talking about *unfurling*, unrolling, and the mere reshaping of stuff in space—the impossible assumption that static substances, with a pinch of movement, shall produce any type of transformation (precisely at the price of reducing all change to spatial reorganization). However, rather than *unrolling*, the world seems to be *rolling upon* itself.[10] By unrolling, everything is there, and nothing is new. By rolling upon itself, nothing lasts but nothing is lost. This is the great temporal asymmetry that Bergson does not fail to insist on. That is, the past is absolutely conserved, and the future is absolutely new. Evolution then acquires substantially different meanings. Distinct juxtaposition of substances cannot account for continuous interpenetration of processes. Tragically (yet fortunately for our everyday life functioning as practical beings) the intellect has huge difficulties in thinking the latter and feels at home with the former. We suffer from acute and chronic *misplaced concreteness*.

So, again we ask: how does *the new* ever come into the world? Whence variation?[11] In the movement of explanation part of the answer is to be found.

EXPLANATION IN PRACTICE

Explanation is intelligence's Procrustean bed. Ultimately, an essential presupposition of the rational mind is that any process under scrutiny should be deemed rational at first and for as long as it is not demonstrated to be otherwise. Any suggestion about the non-intelligibility of something is immediately met with protest and accusation. Yet, how can one think about a problem and, regardless of the apparent success in its provisional solution, discover the very same limits of the process of thinking in the process of thinking itself? It is like trying to perceive your own retina. How does one actually see a blind spot? Note the difference between looking at the world through a window and looking at the glass of the window (or seeing our own reflection in the glass, which could in turn be scratched, dirty or tarnished). Clarity of thought can reflect its own opacity.

A second notable feature of explanatory work is that it is both late and early. Being actually in the future with respect to the event to be explained, explanation sneaks back to the past, all the way before the event, to somehow virtually re-produce it after it happened as if it had not yet happened. After the fact, explanation imagines itself contemporary with the fact and, at the same time, explanation shams the fact to finally dispense with it. Explanation is then back to the future.

Explanation is, in essence, a flawed but convenient act of substitution.[12] One could rightly say that all illusions of the intellect build upon substitutions. At the core of explaining evolutionary processes, we are forced to perform a double operation. We make them disappear first in order to make them appear later, in a new form sanctioned by our intellect. Reconstruction implies rejection and postulation; substitution of what was there with what our mind has derived. Yet, the intellect, in its commendable effort, forgets a crucial issue: diversity was there before humans even dared to try to think about it. For instance, I enter in a room and claim it is untidy, but disorder belongs less to the room and more to my frustrated expectation to see a particular type of order. I then protest to my own inability to embrace an order which does not meet my needs for action. There is evolution, and *then* there is our theory of evolution.

When biology, the logic of the living, is reduced to logic, it ceases to be a word about the living. The logic of the argument will substitute the *logos* of Nature; what it has of logic it lacks of living. To analyze, paraphrasing Bergson, then becomes to express something as a function of what it is not. The intellect, of course, needs to be oblivious to its own tricks. And when it is not, it is necessarily defensive. (Yet, let us not conflate the critique of the intellect with anti-intellectualism or some sort of rational defense of irrationality!—actually, if one pays sincere attention, what seems to be going on when the issue is raised is more like an irrational defense of rationality.)

Note how easily we derive the conditions whose absence we try to observe. Not only do we practice the fallacy of considering present what is past (i.e., the phenomenon to be explained) and past what is future (i.e., our explanation of it), but also that of substituting being with non-being, and vice versa. For example, in neuroscience, we think, and then we think how to demonstrate that we think without assuming thinking as a premise. In biology, we live (painfully seeing our pets, friends, and relatives die) and then we figure out ways to derive the living from the inert. We are conscious, and yet we spend decades and billions to explain away consciousness, ignoring that explaining is a conscious act, and neglecting that consciousness—being the very possibility of phenomenality—cannot be reduced to an epiphenomenon, which is actually an abstraction (say, neurons firing). We take metaphors for mechanisms, and, even worse, we forget that a mechanism is also a metaphor.

These are the procedures of the diminution fallacy. Being sort of a "demiurgic prejudice," the assumption is that in its centripetal movement, the intellect tries to account for the centrifugal movement of life; to go from less to more; from parts to whole.[13] In other words, how shall we understand something hoping to start from so little that it practically amounts to nothing? "Does it not in fact lead to something because it furtively inserts something into its nothingness?"[14] Occam's razor, having nothing to cut, has cut our throat. The misplaced notion that "more is different" when one piles up "more of the same" is at the base for the hypes and hopes of scientific reductionism. This is an imperative of the intellect; it is unemployed otherwise. This movement of fabrication is labor and progress thought in terms of scale. It is the substitution of the scientific spirit by engineering coupled with the technician's idea; namely, the use of linear automated work to amplify its effects quantitatively. This is the kind of intellectual effort that differentiates authors from copyists.

To cut to then sew; to empty to then refill; to halt to then resume—these are amongst the main habits of the intellect. What can be put together can be split apart, but the opposite process is not necessarily true. What is a plenum must be conceived as a plenum, and so it cannot be built piecemeal. Out of nothing, nothing comes, but one can perform the double operation of thinking about something and then imagining its absence. As Bergson noted, nothing, in this sense, is literally more than something. One can go from mobility to immobility via diminution but not the opposite (as Achilles' tortoise attests). Intelligence cannot lead by augmentation to *intuition*.[15] The gap requires a leap.

To conflate fact and abstraction has major consequences here.

> These two logical characteristics of mutual distinction of terms and externality of relations certainly do belong to the abstractions employed in explanations, and we commonly suppose that they belong to everything else besides. Bergson, however, believes that these logical characteristics really only belong to abstractions and are not discovered in facts but are imposed upon them by our intellectual bias, in the sense that we take it for granted that the facts which we know directly must have the same form as the abstraction which serve to explain them.[16]

The novelty produced by the intellect is not really new. No feat of human abstraction based on a recombination of tokens lends itself to conceiving true creative advance. Only immediate experience (and perhaps imagination) can.[17]

If the universe is not truly novel right now, then it never was and it will never be. Again, we must ask: whence *novelty*?[18] And so we arrive at the problem of *continuity*, and its two flavors.

DURATION AND HETEROGENEOUS CONTINUITY

The intellect, at home with space, makes continuity discontinuous at will, and thus dull.

Let us now see the difference between homogeneous discontinuity and heterogeneous continuity. In order to understand evolution (*and what our intellect makes of it*), it is imperative to say what *duration* is:

> Pure duration is the form which the succession of our conscious states assumes when our ego lets itself live, when it refrains from separating its present state from its former states. [...] it is enough

that, in recalling these states, *it does not set them alongside its actual state as one point alongside another*, but forms both the past and the present states into an organic whole, as happens when we recall the notes of a tune, melting so to speak, into one another. Might it not be said that, even if these notes succeed one another, yet we perceive them *in one another*, and that their totality may be compared to a living being whose parts, although distinct, *permeate one another* just because they are so closely connected? [...] We can thus conceive of succession without distinction, and think of it as a *mutual penetration*, an interconnection and organization of elements, each one of which represents the whole, and cannot be distinguished or isolated from it except by abstract thought.[19]

The notion of Bergsonian *durée* (duration) allows the mutual accommodation of continuity and discontinuity. The antinomy can be surpassed by the fecund and new distinction between time and space. Then we see two types of multiplicities: a quantitative one and a qualitative one. The former is based on *juxtaposition* in space, the latter on *interpenetration* in time. What is homogeneous, clear, and (because of that clarity) distinct can only be in space; what is heterogeneous, precise, and (because of that precision) con-*fused* can only be in time. A sacrifice must be made by our intellect (ironically, the analytical philosopher would demand a clear definition of precision, rather than attempt a precise notion of clarity). Arguably, its most productive habit—namely, the systematic spatialization of what is not spatial—must be dissolved in favor of the primacy of our experience, which proclaims otherwise; the realization that time in space is not real time: "This indivisible continuity of change is precisely what constitutes true duration."[20]

The discussion about continuity and discontinuity requires one to grasp the essential difference between time and space: "Instead of a discontinuity of moments replacing one another in an infinitely divided time, it will perceive the continuous fluidity of real time which flows along, indivisible."[21] In other words, evolution, thought of as unfurling, even if blind, betrays a non-spatialized notion of time. "Without the continual unfolding, there would be only space, and a space that, no longer subtending a duration, would no longer represent time."[22] The same essential distinction is implied in continuity and discontinuity. In a conference on the psychological origin of our belief in the law of causality, Bergson maintained that "la différence entre les idées de continuité et de discontinuité est une différence essentielle. *La philosophie de la contingence signifie que tout ne se ramène pas a une différence de mécanisme, d'arrangement.*"[23]

As Spyridon A. Koutroufinis says in his chapter for this volume, "[t]he duration of our experience constitutes a heterogeneous continuum." It is

> the stream of interconnected experiential qualities that permeate one another. These mental acts determine their own essence through their interpenetration. The whole *duration* is a process that permanently transforms its own essence. Thus, it is in the most radical sense of the word a heterogeneous multiplicity. In sharp contrast to the homogeneous continua constructed by abstract intelligence,

such as the empty three-dimensional Euclidean space of Newtonian physics and the mathematical continuum of numbers,

> it constitutes the most concrete continuum that we know: because it incessantly transforms its own essence it cannot be subject to any mathematical operation.[24]

Note that I am not claiming one cannot understand evolutionary theory (nor that one could not understand evolution in theory) but that one does not and cannot truly understand evolution itself by means of the spatialized notions of the intellect.

Our outsourced experience of time—our thinking about it—makes it homogeneous. *Duration* is heterogeneity. Abstract space-time is homogeneity. The main intuition here is that concrete psychological continuity is always heterogeneous. Then if, in order to have heterogeneity, we force a discontinuity,[25] we must recur to homogeneity. But, if we inject a discontinuity (psychologically and epistemologically), homogeneity seems unapt for the job.[26] We are led to contradiction one way or another. One can always zoom in to that pseudo-time (abstracted time, made space), searching for the curve smoothly ranging from A to B.[27] It is not casual that we commonly use the expression "to take place" to signify that an event occurred (in time!). The intellect, then, must pay the price of freedom: one is compelled to choose between continuous or discontinuous. A selfish (and sterile) sacrifice, indeed, because the intellect will not propose itself for the offering. Evolution seems to be continuous in itself—a heterogeneous continuity, which we may think of as discontinuous. Rather than a chicken and egg problem, the fish ate its tail.

Homogeneous discontinuity[28] is a surrogate for heterogeneous continuity. We may talk of two essentially different forms of continuity and of discontinuity. Bergson's notion of continuity is rooted in the psychological, the heterogeneous manifold of pure experience. In this context "heterogeneity" means permanent new creation of a multiplicity's or manifold's essence. Something continued persists in activity, while something continuous forms an unbroken whole. "Creation would have appeared not simply as continued, but also as continuous."[29] The second kind of continuity, the abstract/mathematical kind, is based on the homogeneous manifold of geometric space and numbers. In this context "homogeneity" means no change of a multiplicity's or manifold's essence. Bergsonian heterogeneous continuity can be described by the intellect only as a discontinuity because, for the intellect, all continua must be homogeneous in the sense that they do not transform their essence. So, from the perspective of the (spatialized) intellect, a non-homogeneous being must be discontinuous. When it comes to discontinuity, we can conceive interruption of the permanent transformation of essence, like the death of an organism or the destruction of our planet by an asteroid. From a Bergsonian perspective, there is a qualitative transformation of essence—a jump, a leap—between humans and animals, but not a gap. The second kind of discontinuity always presupposes a gap between two abstract homogeneous continua. Because the intellect cannot think[30] in terms of transformation of essence, we see a gap between animals and humans and not a leap. Evolutionary theory, based on homogeneous discontinuity, contradicts itself and evolution.

Our need for tethering defeats its purpose. From immobility, one cannot make sense of transitions. Consider continuity, "[t]he aspect of life that is accessible to our intellect—as indeed to our senses, of which our intellect is the extension—is that which offers a hold to our action. Now, to modify an object, we have to perceive it as divisible and discontinuous."[31] Evolution is a single indivisible history.[32]

Humans are obviously animals, and also more than mere animals. When thought in terms of an abstract homogeneous continuum, we are a discontinuous break with other animals in that we think of evolution. We do so as a discontinuous process, precisely because our intelligence is unable to think heterogeneous continuity[33] (or can only do so by exceeding its normal limits). And this is the whole crux of the matter: continuity and heterogeneity are self-contradictory to the intellect.

On the other hand, our immediate experience of time presents itself as continuous heterogeneity. But my needs make discontinuous my perception of it.[34] Our primary goal is then to "bring together all sensible qualities, restore their relationship, and re-establish among them the continuity broken by our needs."[35] Evolution can then be seen to be continuous and with leaps. Yet, in reflecting about human versus animal, let's keep in mind that, while speciation happens sometimes, evolution "takes place" all the time.[36]

The notion of organism—implicit in Bergson and explicit in Whitehead—can resolve the tension between mechanism and finalism. It does so by abandoning what both share. Regarding the idea of *vital impetus*,

> [w]e have not mentioned, save perhaps by implication, the essential one, namely the impossibility of forecasting the forms which life creates in their entirety by discontinuous leaps, all along the lines of its evolution. Whether you embrace the doctrine of pure mechanism or that of pure finality, in either case the creations of life are supposed to be predetermined, the future being deducible from the present by a calculation, or designed within it as an idea, time being thus unavailing. Pure experience suggests nothing of the sort. 'Neither impulsion nor attraction' seems to be its motto.[37]

Several decades after this quote, it may seem that the polemics between gradualism and punctuated equilibrium are not an issue anymore. Nevertheless, relatively recent efforts to bring a kind of absolute contingency to the modern synthesis do not dispatch the topic. Since all heterogeneous continua permanently transform their own quality or essence they consist only of those creative leaps. They are uninterrupted manifolds of qualitative leaps. Contradiction is dissolved with an intuitive jump. But our linear, causally fixated thinking literally does not follow. It is willing to take as many swimming classes as necessary, always in the solid support that the familiar dry land offers to it, constantly postponing (thus refusing) to just dive into the water.

The implications are not only theoretical. They are personal and ecological. Intelligence, having become its own purpose, is more and more disconnected from nature, up to a point where one could say that, in human beings, everything happens as if nature was deficient. The force is absorbed in itself. Explaining away psychology by physiology is absurd "since the operation consists in destroying the very condition that makes the operation possible."[38] One cannot help but see here a striking analogy with the destruction of life on the planet by humans; the yoga of objectivity requires higher doses of abstraction, which requires distance, and sacrifices our capacity for empathy.

The many flavors of transhumanism have, consciously or unconsciously, looked for an escape. Perhaps concurrent (as a divergent line of evolution) to that of technocratic imagination (and more akin to psychedelic exploration) a reconnection of humans to nature is attainable by the faculty of intuition, and its method. This could bring humans back to nature, precisely by transforming human nature. Or at least its mind. How can we untie the knots of the intellect?

INTUITION: THE MIND UNCHAINED

Intuition is an act of will that requires radical sincerity and, to some extent, an effort of surrender entwined with an intellectual sacrifice. It is then possible to psychologically grasp a type of continuity that is not homogeneous, nor forced to be discontinuous when trying to accommodate its being heterogeneous. Life, not just our inner *duration*, is then revealed as a generic form of such heterogeneous continuity. Duration belongs to life and, as such, it is intrinsic to evolution as well.

Any theory of life is inescapably entangled with a theory of knowledge. We are insisting on the real challenge in evolutionary theory: to overcome its own theoretical limitations. The articulation of an understanding that can grasp evolutionary novelty within an undivided whole has other consequences. It proposes that human beings are indeed an unprecedented leap, yet within the indivisible continuity and multiplicity of creativity that is realized by means of evolution.

The felt presence of immediate experience is the remedy against abstractions. Abstractions pay the price of the concrete. Concrete stuff is de-concretized by being re-expressed, and thus reduced, to universals and their combination. Individuality is universalized and localized. We escape the particular to erect an explanation of it. Our reconstruction of evolution is not evolution itself, in the sense that our abstract reconstructions of evolution embrace the logic of our intelligence but not the logos of evolution. Paraphrasing Bergson, the notions of variation and selection explain evolution rather than produce it; they are not the cause nor the result of the phenomenon, but a part of it. They express it. A static link between the past and the future shall secure the fruits of our action in terms of the choice we gave up. To explain the concrete is an oxymoron. The intellect refuses true creativity because it is unable to think it. There is a sacrifice to be made: an excess of knowledge by the excessive means of the intellect tames will. What is needed is to stand before the phenomenon without the pretense of embracing it in order to control it.

Let us insist on these vices of thought, as pointed out by Bergson. *We pretend we go from less to more.* Therefore, we actually need to pre-empt the plenum. This is a double operation: from whole to part, and then reconstruction. But parts are always partial views. In a similar fashion, *the real is made subordinate to the possible.* We believe that something actually happens because it was possible in the first place. Yet, the possible does not precede the real (unless by our imposition; once something is real we say it is so because it was possible). The intellect seems to enjoy veto power over the real. *We think the void is less than the whole.* Nevertheless, the whole is primary, and, on top of it, we add a second psychological operation that negates it. Disorder presupposes order that is then lost. Nothingness presupposes a whole that is then removed. Discontinuity presupposes continuity, then divided. Continuity cannot lend itself to divisions that freeze it. Becoming must be undivided, even if it remains divisible.[39] (This is what Nietzsche defined as our sense of loss, because we are fixated on the past of what was present, and we know change as negative being, but this is a wrong perception of time.)

As in evolutionary biology, so in neuroscience. Such a fallacious *modus operandi* cannot resist applying itself to every single topic. The explanation of mind-like phenomena (let us take consciousness as its major exponent) either negates the phenomenon, explains it away or, at best, sees it as an emergent process from the unconscious, rather than a fundamental principle. The unconscious, being defective in consciousness, is postulated as primary. But, any self-dormant manifestation presupposes the sleeping beauty. If the movement of molecules can create feeling with a nought of consciousness, why may consciousness not be able to create, in turn, movement with a nought of energy, or simply use it as it pleases? For Whitehead and Bergson, creativity is the ultimate process. A *why without because.*

Therefore, as a rule of thumb, when forced to choose between the two alternatives (continuity versus discontinuity), we better reject the negation, since in order to negate, an affirmation needs to be there first.

> To this possibility of decomposing matter as much as we please, and in any way we please, we allude when we speak of the continuity of material extension; but this continuity, as we see it, is nothing

else but our ability, an ability that matter allows to us to choose the mode of discontinuity we shall find in it. It is always, in fact, the mode of discontinuity once chosen that appears to us as the actually real one and that which fixes our attention, just because it rules our action. Thus discontinuity is thought for itself; it is thinkable in itself; we form an idea of it by a positive act of our mind; *while the intellectual representation of continuity is negative, being, at bottom, only the refusal of our mind,* before any actually given system of decomposition, to regard it as the only possible one. *Of the discontinuous alone does the intellect form a clear idea.*[40]

Since the foundations of Western thought, stability has been rated higher than change.[41] As it seems that a theoretical step forward of Darwinism was to replace reification (stability) by change (evolution), the preference for substance over process is an old ingrained habit of our Western minds that remained untouched with neo-Darwinism. As a consequence, we conceive continuity as indifference to the discontinuity that our intellect imposes on life via the notion of juxtaposition in space.

Nearly all the sciences (and a great deal of philosophy) face a strong philosophical caveat, one could even say an oxymoron: to try to explain what is complex by means of what is simply a sprout of the retrospective illusion, since what is complex is complex *de facto*, while what is simple is simple *de jure*. The habit of transforming the description of a phenomenon into its prescription is well ingrained. Newton's laws, explaining the movement of terrestrial and celestial bodies, seem to actually produce it too. Because we can describe the process of evolution in terms such as trial and error, we then postulate that Nature is operating on those terms. The phenomenological approach—in its light version, a kind of skepticism without cynicism—is thus an honest beginning to untie the knots. Since avoiding projection seems impossible, one should consciously try to recognize what is projected, and to avoid imposing that as a cause of what is to be explained. One can safely speculate that neither the so-called love for truth that is meant to define the activity of philosophers, nor the explanation of the real by means of the study of the natural world via observation and experiment that occupies the time of scientists are actually innocent efforts. Underneath—and actually often bluntly erected at the surface—one finds an intention to make use of truth and the natural world. Disinterested knowledge and curiosity so quickly give way to profit and utility, and then creativity is gone.[42]

We are back to the question of *the new*, now from the perspective of *difference*.

A LOGICAL DISTANCE

The intellect is the great homogenizer. For it, the production of difference can only be the occurrence of recombination. For it, nothing is really new in time. For it, all is distinguishable only in space. Spatialized things are distinguishable because each of them occupies a separated piece of space, having thus very clear borders. The intellect draws clear borders around all things, real and abstract (namely, concepts). Heterogeneous continuity is truncated into homogeneity followed by discontinuity. In a word, what we want to clarify, we must make imprecise.[43]

Within the conditions for the intelligibility of the real, there is, of course, the principle of difference: to find what is different and what is the same. Intelligence (etymologically, to read inside) has gone from seizing the thing inside (nearly knowing by identity) to literally seeing its parts (analysis at a distance). In their difference, what happens is what they have in common: something interchangeable. The principle of difference is at work after conflating similarity with sameness. Groups, made upon similarity, disregard the very concreteness that makes things. They can then be compared according to their difference in what they are not. When the same goes with the same, we can talk about repetitions.

The path of evolution points in that direction too. The stability of matter is a guarantee that processes endure long enough to be called things, thus being the durable base of change (you never cross the same river twice, yet tables last long enough for me to be able to write these lines about the river). Then repetition, a vestigial trick of life to convince matter to give up part of its stability, issues recurrence as a promissory note: an engine employed to perpetuate structures while opening the doors of creativity. Life strives to make room for difference within the processes of repetition. As a characteristic power of the intellect, difference abstracts facts to make them stable and then compares them systematically. A triple nonorganic operation: undressing, freezing and contrasting. Yet,

> [w]ere events unceasingly mindful of their own course, there would be no coincidences, no conjunctures and no circular series; everything would evolve and progress continuously. And were all men always attentive to life, were we constantly keeping in touch

with others as well as with ourselves, nothing within us would ever appear as due to the working of strings or springs.[44]

There is never real repetition, just pure creativity, or at best "the repeatable gradient of the unique."[45] The maxim is a reversal of Nietzsche's: nothing lasts (pure creativity as a principle) yet nothing is lost (the whole past is in the present[46]).

Difference is perhaps the most important endpoint of the workings of the intellect. We excel at the classification of phenomena into different boxes. We enjoy rejection of the null-hypothesis via statistically significant differences. Take the common expression "to make a difference" embedded in our language that equates differences with to have a significant effect. To what extent can a concentration on sameness make a difference? The etymology of difference, to carry away, indeed brings us afar from the thing. Distinction maintains a distance. It puts apart. To discern is to separate. But not only do we pull things apart; we blur them in order to do so. And in order to separate, we need to express the concrete in terms of what it is not (a re-presentation, a symbol, ultimately: an idol). Only then can we label the box where our intellect will put things. Keeping a distance is a characteristic attitude of the intellect: I explain what I see I cannot touch anymore. Contact is gone, and most of our job is to re-establish it.[47]

A WILLING PROXIMITY

Is there thought beyond thought? While intuition is a higher potency of will, abstraction makes it weak.[48] Life processes transform their own essence (recall the notion of heterogeneous continuity as a permanent transformation of essence). Intuition is a higher vitality, based on attention,[49] a continuously moving attention.[50] It is able to grasp the concrete without the double operation of forcing similarity on it in order to extract a difference. To the intellect—whose favorite plant-based substances are caffeine to ramp up (morning "coffee breaks") and alcohol to loosen up (evening "beer hours")—heterogeneous continuity is neither homogeneous nor discontinuous. Let's say two things are the same. What else is left to say about them? Now, if they are similar, how can these two things be compared if not by putting them onto the same score via a token that annuls what is concrete in them and extracts from them what does not really belong to them?

The path to difference taken by the intellect performs artificial operations of sameness. The path of sameness taken by intuition seizes directly what is uniquely different in everything. The static sameness of the mathematical equality sign is conservative; the inequality sign brings disconnect. The distance between two objects translates nothing about them except that which is not properly theirs.

The need for distance implies a critical reduction of empathy. The quest for clarity, both in analytic philosophy and mainstream science, demands to explicate order of any degree as order of lowest degree. Searching for precision, rather than clarity, does the opposite; it accepts a certain nebula in order to accommodate things as they are. In other words, the intellect refuses to be touched by things, while intuition dives into the things themselves. The fact that intellect then protests that such intuition is a chimera, something vague and made up, only reinforces this point: that intellect is unable to think in proximity.

This is anathema. Biology requires a theoretical base not deducible from the fundamental concepts of physics and chemistry: "For the biologist, the world of the physicist has only the value of a world created by thought; such a world corresponds to no reality, but it is to be considered as indispensable aid to calculation."[51] In the decades after the first English publication of Jakob von Uexküll's *Theoretical Biology* in 1926, the discipline progressively turned into mathematical biology. The original intuitions that animated it became spatialized, frozen by the intellect. A true biologist cannot solely be a logical artisan ("Know Thyself"). As Uexküll wrote: "a preliminary condition for the investigation of the appearance-world of others is an exact knowledge of our own."[52]

When concerned with the process of evolution itself, the door stays locked to any sort of teleology whatsoever; the explanatory power of the intellect wants to get there by a linear chain of "push-push" operations, reluctant to any "pull." It is not our intention to open that door either. We will have something more to say about teleology and mechanism later.

The accusation thrown at vitalism is hardly more than an automatic mockery built upon a misunderstanding of Bergson's élan vital. Such impetus is the unbroken force of creativity:

> The cause of growing old must lie deeper. We hold that there is unbroken continuity between the evolution of the embryo and that of the complete organism. The impetus which causes a living being to grow larger, to develop and to age, is the same that has

caused it to pass through the phases of the embryonic life. The development of the embryo is a perpetual change of form. Any one who attempts to note all its successive aspects becomes lost in infinity, as is inevitable in dealing with a continuum.[53]

David Lapoujade expresses the effort and ability to reach such impetus of life in thinking:

it is not anymore an observing from outside, it is to the contrary to enter into the interior of another point of view, as one sympathizes.... Bergson opposes those who have a point of view about the movement of a phenomenon with those who make of movement the very point of view of the phenomenon, their "consciousness." Then for instance, mechanicism, finalism, evolutionism have a fixed point of view about the evolution of species while Bergson makes an effort to reach the "consciousness" or inner impulse of the evolutionary movement.... And since everything is movement, is the whole moving reality what develops a point of view of its own.[54]

Excelling in the art of the physically possible, the physical sciences (obliviously backed-up with a worn-out substance metaphysics) projected, with reasonable insistence (often turned into a stubbornness that defeats its own purpose), onto the phenomenon of life the same immobile scaffold of the inert. And it now requires for its own justification a Procrustian adjustment of the phenomenon it attends to. The inability of our intellect to conceive intuition and to think intuitively is not an argument against intuition.

Science campaigns with zest that variability has nothing to do with supra-natural forces anymore. This is understandable, and it probably was necessary. But what is often meant by super-natural or supra-natural is that with which the rational-logic mind is not at ease (our reaction to it due to our own inability to think it). Or, more bluntly, the natural is equated with the intellect since the intellect, following the Baconian dream, thinks its job is to control nature. *I can, therefore I must.* And so, whatever does not lend itself to that endeavor is immediately suspicious of being unreal and impossible (an illusion, woo-woo, bla-bla, etc). The conviction that what the intellect cannot embrace ought to be a superstition is itself a superstition. Then, what cannot be rationalized is accused of being supra-rational (thus immediately seeking death-row pardon), and, ultimately, explained away; e.g., reason as the measure of all things, including nature; too much pride after an excess of humility.

And so, the copy-and-paste myth soon wonders how to sew the cloth of life. Recombining abstractions seems to give rise to variation and the new. And the new is only thought because it is actually not thought. Abstracted, and by definition external to each other, is such human-made operation the only principle we are capable of in order to reassemble the whole or, better, recover the initial contact with it?

SPONTANEITY TAKEN SERIOUSLY

Duration is a continuum of transformation of its own essence. As such, it can only be a heterogeneous continuum. Whitehead's *concrescence* and Bergson's *duration* are modes of expressing the self-determination of essence, the notion that that of which the world is made has the intrinsic ability to transform itself. This vividly confronts "thought life" and "lived life" and, correspondingly, "thought's necessity" versus "life's necessity": "La nécessité inhérente a la loi de causalité se déplace ainsi entre deux limites extrêmes: de *nécessite vécue* elle devient *nécessite pensée*. Empirisme et apriorisme s'accordent, au fond, a ne tenir compte que de la second de ces deux formes de la nécessité."[55]

Bergson and Whitehead modify our understanding of causality inherited from classical physics according to which the cause (past) determines the effect (present). When relations are made external through cause and effect, then their immanence is made incomprehensible. But causality, used in the loose sense of connection between events that are not totally arbitrary, can be adapted to process philosophy.[56] If actual occasion A prehends actual occasions B, C, D, and E, then those occasions are the efficient cause of A. But they do not determine A in the sense that the prehended entities do not determine how A will include their essence in its own self-determination. In a word, A decides through its own self-determination how the efficient cause will be integrated in the process of A's self-formation. A determines its own essence.

The process selects its causes. In process philosophy the prehending entity selects which parts of its past will be integrated in its self-formation and also decides how this integration will be formed. In Whitehead's philosophy of organism, each actual entity is to a high degree the cause of itself; a self-forming, self-determining actual occasion is to some degree its own cause, *causa sui*. We read in Whitehead's categories of explanation that "actual entities are the only reasons; so that to search for a reason is to search for one or more actual entities."[57] This principle of efficient and final causation is termed the ontological principle, where final causation is "the internal principle of unrest," whereas efficient causation "is a ground of obligation characterizing the creativity."[58] Processes are self-caused, and this is very difficult for the intellect to conceive. In a "weak sense," some attempts in theoretical biology have made progress in understanding biological autonomy. Yet, in its "strong sense," à la Whitehead's *Process and Reality,* creativity is the ultimate principle.

Let us make a political digression. Science, philosophy, and philosophy of science become academic shell games if their political dimension is suspended. Note how easily defending one's own salary is conflated with defending the sanctioned ideology... There is an easy slip from postulating evolution as a blind mechanism of struggle so as to justify a political position, tainted with neo-liberalism, that sanctions the idea that any social encounter is primarily an act of competition. When A and B are related externally, their relation is easily cast as competitive, and so that relation must lend itself to quantification for subsequent comparison. They enter directly into disjunction: *either A or B, rather than both A and B.* Internal relations, instead, make both A and B essential to each other, and so their conjunction is primary. (The symbiotic view of evolution conceives the relation between two entities as collaborative, rather than competitive; yet those entities may or may not be internally related.) We arrive at two hard-swallows for the intellect: continuity of mutual interpenetration of essence and self-determination.[59]

Indeed, the world bubbles forth, as Heraclitus famously said. Everything flows. We cannot step into the same river twice. Yet, he also said that all is the ever-living fire; night and day, good and evil, are One. Space is a principle of differentiation, while duration is the continuous elaboration of the absolutely new. Life, in its divergent lines, is continuous.

Let us not take Bergson's words as simple allegories, but as an accurate description of his experience: "the continuous creation of unforeseeable novelty which seems to be going on in the universe."[60] Then, why don't we see it? In discussing continuity and discontinuities in evolution one may decide to discuss nature but not us, or else make the effort to realize that our own thought determines what we see in nature and ourselves. If we treat time as space, then time has no duration. Real time does not lend itself to juxtaposition. Events are not lined up, one next to the other, in space. Quantitative multiplicity, in order to count, needs to enumerate in space. Qualitative multiplicity allows for heterogeneity without juxtaposition.

"Because a qualitative multiplicity is heterogeneous and yet interpenetrating, it cannot be adequately represented by a symbol; indeed, for Bergson, a qualitative multiplicity is inexpressible."[61] A qualitative multiplicity is therefore heterogeneous (or singularized), continuous (or interpenetrating), oppositional (or dualistic) at the extremes, and progressive (or temporal, an irreversible flow, which is not given all at once). Pure duration excludes all idea of juxtaposition, reciprocal exteriority, extension and indivisibility. Evolution can be understood without being explained.

CREATIVITY

The psychological point of view from which *duration* is grasped can then be applied to life (and also to matter, as opposed to the inverse route of starting from physics to derive biology and psychology[62]). The faulty over-emphases of mechanism ("push") and finalism ("pull") can be overcome:

> *the psychical life* is neither unity nor multiplicity, that it *transcends both the mechanical and the intellectual, mechanism and finalism* having meaning only where there is "distinct multiplicity," "spatiality," and consequently assemblage of pre-existing parts: "real duration" signifies both undivided continuity and creation. In the present work we apply these same ideas to life in general, regarded, moreover, itself from the psychological point of view.[63]

Thought thinks a universe that operates, while experience feels a universe that creates. The notion of mechanism precludes the possibility of real change because the future would always be contained in the past (or else indifferent to it, in a-causal accounts). Note that, etymologically, to produce (take as "to cause" efficiently) is to lead forward. This can be conceived either by thrust or by attraction.[64] Both for the mechanistic and finalistic views, all is given at once. What changes is from where the tape is unwound. Creativity is then impossible. Yet, processual finalism is different: the end (aim and purpose) is not fixed during the process but emerges in an unforeseeable way in the process of Bergsonian *duration* or Whiteheadian *concrescence*.[65] All is new all the time. The finalistic element is not a demiurgic planner but a striving for actualization and self-determination. But the intellect prefers to think of the unpredictable as the temporarily unpredicted.[66]

Pushed from the past or pulled from the future, in neither case is explanation contemporary to the phenomenon to be explained. In other words, whether the cause preceded the effect, or the effect preceded the cause, behavior is never conceived as present participle. Being in conflict, both mechanism and teleology make the same inadequate assumptions. After-telling—sold as pre-telling—is never how things were. The dance between the normative and the conditional permeates explanatory work. Things should have been this way; things could have been that way.... Yet the dialogue between *possibility and necessity* falls short in capturing how things actually are.

Our thought acts as if there were a kind of surplus in the real (while, ironically, and as we have exposed above, it pretends to go from less to more by means of the application of "thought necessity"). Any model implies compression, which is the idea that we can leave out a part of reality, due to its redundancy in theory, in practice, or both. But, why would something be a superfluous happening in nature? And, superfluous to whom? for what?

We happily take forward movement as backwards:

> As I cannot predict what is going to happen, I quite realize that I do not know it; *but I foresee that I am going to have known it*, in the sense that I shall recognize it when I shall perceive it; and this recognition to come, which I feel inevitable on account of the rush of my faculty of recognizing, exercises in advance a retroactive effect on my present, placing me in the strange position of a person who feels he knows what he knows he does not know.[67]

The retrospective illusion, of course, can be turned into the prospective illusion: the belief that in order to know what will happen in the future one simply needs to study in great detail what happened in the past. Such an assumption is based on the double movement of rewinding and fast-forwarding the tape of reality. It is quite a Western thing to try to accelerate and decelerate reality at will.

Mechanistic reductionism presupposes what it claims to explain. The movement of decomposition by reduction to lower levels is a retrospective accounting. After having found the so-called neural basis of thought or the genetic basis of evolution, thought and evolution are claimed to emerge from them (and subsequently ignored). Other koans from the same family of fallacies then read as follows: thought abstracts neural mechanisms to think thought; thought abstracts genetic mechanisms to think the evolution of thought. The force of evolution, being the starting point, is presented as a product of the force of thinking.

We tether ourselves to an unchanging conviction. We reassure ourselves that what takes place must be the only thing that could have taken place. We accept the strange idea that an event could have been caused before it has happened. Verbal tenses reflect the fallacy *post-hoc* and *pre-hoc* ("this would not have happened, had I done this or that"). As we said, even when B follows logically from A, it may not do so, biologically or psychologically, but only as reconstructed in our abstracted psyche.

Parsimony, the necessary austerity practiced by science, betrays the concrete in the name of the useful. Note that something can only be useful because it is forced and expected to recur. But being in time requires becoming in time. Instinct and intelligence are bound to survival, whereas intuition is self-transcendent in the sense that it becomes an understanding not restricted by (nor opposed to, perhaps just indifferent to) utility. We shall not blame science if we concede that science does not hold the monopoly on accessing reality, but just that aspect of reality that is abstracted so as to be repeatable, and thus profitable. The unique, the miraculous, are literally anecdotal (etymologically, not worthy of publication). Politically, at the other extreme of pseudoscience, the scientific narrative of our times erects the average scientist as the float bearer in holy processions of progress and statism (Feyerabend comes to mind).

However, the standard procedure in physics, readily exported and commonly practiced in biology, decrees that what is a result must act as its own origin; that "the endpoint ought to be the point of departure." This "consists in leaving what is in the making, in placing oneself after the fact, and in performing, a posteriori, a little justificatory reconstruction thanks to which belated abstractions become primitive only because they are simple and poor."[68] The intellect relaxes. Such released tension of the mind gives rise to extension, namely, space. Our action can take place. Extracting one side of the polarity, a medium is born. This medium is one that tolerates discontinuity—even more, it celebrates it, since the very possibility of divisions lies in its indifference to cuts. The homogeneous is born by diminution of the heterogeneous. Space is homogeneous, yet time is continuous; a continuity pregnant with difference! Such difference, having divisions, does not lend itself to division by our intellect without altering its essence. It looks as if time were divisible but undivided. Time is a continuous heterogeneity, which means that it is a unity of plurals. The fact that we can think in terms of abstractions clearly implies that reality lends itself to abstraction, yet it does not permit a full elimination of duration. Every explanation then consists in dealing with the problem at hand in terms of space. When time itself is dealt with in terms of space—which is the essence of Bergson's critique, and also his great insight—the many problems (specially those concerning the life and mind) become unsolvable due to an inherent contradiction at the core of our perception. Time is not a succession; it is an accumulation, a growth. The path is creative. Everything is always new—the path endures; nothing is ever lost. Life is in time more than space. It is not only more creative than we suppose but more creative than we *can* suppose. The remaining outstanding question is always the same: a proper psychological critique of the space-time concepts.

Biology—the logic of the living—is the precise terrain where the battle between physics and psychology takes place. The movement of mind confronts the inertia of matter. Purpose and will undergo the psychedelic trip of seeing their reflection as dull force and blind activity. Mathematics cannot solve the opposition between determinism and free will.[69] If the universe died and was reborn at every instant, would we be able to tell? Biology struggles to solve the opposition between life and matter.[70] Neuroscience faces the same problem with respect to the body and mind problem. Duration has to be irreversible. For the physicist, the arrow of time manifests itself in thermodynamics. For the biologist,[71] the arrow of time is obvious in the life and death of the organism, and in the life of the species through evolution. For the psychologist, the arrow lies in the very same stream of conscious experience.[72] A living being is a center of action. Choice—which is not a decision between possible alternatives, precisely because one only exists after the act (and the rest never did)—is creation, and creation is labor. Labor is an effort different from work. One could say that work is the effort of intelligence, whereas labor is the activity of intuition.[73]

But we read, and we think about what we read (or rather we read barely more than what we think), and then, still, if we are honest, it seems that change cannot be real change. What we hear is not what we see; our eyes make our ears deaf. Our tongue is silent. Do you see what I say? Can you say what you hear?

The intellect resists the immediate data and postulates re-arrangement as the only possible source of change.

> The metaphysician that we each carry unconsciously within us, and the presence of which is explained, as we shall see later on, by the very place that man occupies amongst the living beings, has

its fixed requirements, its ready-made explanations, its irreducible propositions: all unite in denying concrete duration. Change *must* be reducible to an arrangement or rearrangement of parts; the irreversibility of time *must* be an appearance relative to our ignorance; the impossibility of turning back *must* be only the inability of man to put things in place again.[74]

To test the pure aspirations of the scientist and the philosopher, one may confront them with the following "shock or death" proposition: what if what you wish to understand does not lend itself to analysis? The philosophy of duration-organism lends itself as a powerful ontology for twenty-first-century scientists. One must debate between campaigning for it and patiently awaiting for the Cartesian orgy to lose interest. Would we keep the hammer and toss what is not nail-like? Can we see the limits of intellect with our intellect? Isn't it a dangerous conflict of interest to ask my intellect for a self-report on its limits? I do not mean something hard for it to do but the confession that certain questions about our experience of reality are just not suitable to the intellect's own operations and essence. If the intellect declares something as impossible, would our will still deem it worth knowing? Is the kind of thought that is able "to think matter" (to swiftly manipulate it) the highest point of knowledge? This indeed sounds like bad news for science in general and for biology in particular, especially when bio-logos means the application of the logic of matter-intellect to life. Intelligence, as any structure of dominance, kicks or kills whatever threatens its supremacy, if not its very existence. Intelligence cannot think actual occasions. Bergson and Whitehead, thus, can't be mainstream in academia, even less so in Western minds. Intelligence thrives in outwardness. It needs distance. Intuition, in turn, is the effort of proximity. It knows via inwardness. The derived cannot create the given. Let's put it bluntly: *for intelligence it is impossible to think that something creates itself.*

Paradoxically, symbol use is then both what made us different from animals and what keeps us from becoming fully human. And so, as stated by Spyridon A. Koutroufinis, whereas the evolution of "human intuition will further divide humans and animals, it will open human beings to unprecedented and unforeseeable ways of empathizing with other living beings" and thus with animal life as well.[75] Only then will we become able to empathize and sympathize with animals, plants, and the planet. We must move further away from the animals in order to understand them. Symbols (the tools for understanding mobility by means of halts) represent a progress in evolution, which in turn precludes our understanding of it. Could a conscious evolution take place with symbolic thought?

EVOLUTION'S FOURTH WAY

After having tried creationism, neo-Darwinism, and dynamical systems theory, the intellect must open itself to intuition. The flavor of immanent evolutionism postulated by neo-Darwinism is just too coarse and poor. Such naturalization of human beings was extreme and naive in that it ignored the meaning of natural as projected by us, human beings. The atom has no consciousness, the stone has no creativity, the wind does not show any spontaneity, therefore flies, plants, and birds cannot either.[76] Bergson's, despite all the fuzz against his *vitalism*, is not a proposal for a transcendental evolutionism.

Darwin's main idea is that intelligent design must give way to natural selection. From such a vision, an over-correction (a justified reaction with an unjustified emphasis) proclaims: "No more design than the direction blowing in the wind!" Creationism opposes neo-Darwinism as divine purpose opposes pure luck. Those who (rightly) support the theory of evolution reject heterogeneous continuity; and those who (wrongly) reject it, are more prone to accept it. Natural Selection, the profane god of Neo-Darwinism, has faced the unscientific claims of Creationism with the irony that it cannot still account for the creative Force in nature. It is not an intelligent design versus biological evolution debate (which recalls futile science versus religion struggles). The terrain for quest is in the vertical axis: the intelligence versus intuition problem, namely, how to conceive a sort of continuity that is not dull.

Having been thinking about clocks and screws for so long, when we look at a flower or a beehive, all we can see are little clocks and little screws. The gospel reads as follows: "DNA, the secret of life." If nature is a machine (so is our brain), then we are machine-like. But, is nature a machine? Nature—we insist—must be dispassionate, to the same extent that our study of it should be. Or is it that we chose, for some reason we cannot recall, to be dispassionate in our study of nature and thus could not allow nature to show its passion. We impose a sort of repetition in high-dimensional space, ignoring the true creative force of nature. We take for mechanic what is organic. We deem inertial and reactive what is alive and spontaneous. We search for causes (or lack thereof) to that which is self-caused. The Greek Logos is now barely more than a scientistic Lego. A game, a puzzle, with the hype of neoliberal technocracy.

Attention to abstractions, when exaggerated and excessively sustained, has consequences on our sensibility. Intellect, when practiced at the expense of intuition, provokes atrophy of empathy. Clarity is deemed as the great virtue of thought. Yet, empathy allows for precision. The spherical cow can't walk or produce milk. The inner task of philosophy, so often lost even amongst armchair academic philosophers, is the art of intellectual sympathy. Ironically, critique, which is its opposite, is the over-practiced pattern.

It is then urgent to move beyond mechanistic biology. The arc of progress seems to be from molecular mechanism to dynamical system to organic process. Yet, a re-elaboration of dynamical systems theory cannot capture the essence of process ontology. The former tends towards the latter, but the former also misrepresents the latter. Postulating itself as a means precludes its goal. The term "dynamical process" seems to me more problematic if it is used in a process philosophical context à la Whitehead and Bergson. But, what do scientists and analytic philosophers know about Bergson's critique of spatialization? What is really understood about Whitehead's notion of process as an entity that neither moves nor changes? What does intuitive thinking entail? Thought is most of the time rational at best. Yet, when thinking becomes an intelligent articulation of deep intuitive insights, then the forms given by verbal intellect never exhaust their initial intuition. Intuition makes the beginning, gives birth to insights. The truth must be known before it can be proven.

The quest for truth has abstracted us from reality and then, after giving up truth for convenience and applicability, our use of reason has effectively ignored reality and re-elaborated life in favor of the practical. Experience is always real. Abstraction tends to truth but, by escaping the particular, commits itself to never reaching truth. Our thoughts are not the measure of the world. Evolution in the era of intuition entails a kind of anarchist intuitivism: people have to free their own intellect—and this cannot be done from the outside.

A human can never be purely an animal. At the same time, it seems that we are not yet fully humans.[77] Perhaps the most remarkable discontinuity between us humans and nonhuman animals is the fact that our intelligence draws a qualitative difference between them. The movement of evolution, and our thought, are enclosed in the circles of the intellect.

Yet, when disjunction becomes conjunction, intuition has left its mark.

Continuity need not be homogeneous, nor heterogeneity discontinuous. The key to holding such contradictions is to see whether there are modes of experience that allow sustaining it. Abstraction is not fit for the job. Our thinking is so prone to committing the double error of excess and defect. We see too much and too little: everything is nothing more than the whirl of atoms. The immediate data of consciousness are erased in thought, and then surrogated by thought. Yet, evolution cannot be explained in a language that would owe nothing to evolution itself.

Speculations about the future scientist imply also speculations about the future human, and thus evolution of mankind. We started this piece by noting the irony that what perhaps makes humans most special in the evolutionary tree is that only humans have a theory of evolution. Yet, as we tried to show, evolution explained is an act of intellect, not evolution itself. Indeed, models are not reality, yet there is an important difference between an asymptotic approximation to facts by models and a fundamental flaw in the capability of intellect to grasp the process of evolution. If evolution participates in the continuous creation of unforeseeable novelty, then humans are an abrupt insertion in the great chain of being, and the most important point in Darwin's teachings has, strangely enough, been overlooked. Humans have not only evolved; we are evolving! "The essential thing is the continuous progress indefinitely pursued, an invisible progress, on which each visible organism rides during the short interval of time given it to live."[78] If we are not to bet on the bio-technocratic version of transhumanism, future humans shall sustain intuition on a level where it can routinely surpass intelligence. This brings us to the second main difference between humans and other animals. We, after a great exercise of abstraction by the intellect, can grasp evolution with intuition. By unmaking the limits the intellect poses (which are its own), knowledge, through will (and grace), becomes force. There is little excuse, as exemplars already exist amongst us.

> Evolution is not finished; reason is not the last word
> or the reasoning animal the supreme figure of Nature.
> ~Sri Aurobindo (1913)

FIVE

Human vs. Animal Relation to the Environment:

A Bergsonian-Whiteheadian Perspective on Uexküll's
Concept of "Umwelt" and Cassirer's "Animal Symbolicum"

Spyridon A. Koutroufinis

UEXKÜLL'S CONCEPT OF UMWELT[1]

IN HIS *THEORETICAL BIOLOGY*, first published in German in 1920, Jakob von Uexküll (1864–1944) developed a highly complex theory that might be described as "biological Kantianism." Starting from Immanuel Kant's thoughts about the nature of space, time, causality, and apperception he suggested a unique approach to most elementary biological concepts such as "organism," "perception," "environment," "evolution," and "adaptation," which radically differs from Darwinism and Neo-Darwinism, as well as from the work of all influential theoretical biologists up to the present time. The concept of "Umwelt" was introduced in Uexküll's famous book *Umwelt und Innenwelt der Tiere* (*Umwelt and Inner World of Animals*) in 1909. There, he makes a clear distinction between the terms "Umgebung"— which I translate as "surroundings"—and "Umwelt."[2] The German word "Umwelt" contains the terms "um" and "Welt," the English translations of which are "around" and "world." In German, however, "Welt" implicitly signifies a manifold that is meaningful to a living being because it lives within it. Features of the physical surroundings that are relevant to the organism with respect to its self-preservation and reproduction constitute its Umwelt. An organism will incorporate those relevant aspects of its surroundings into its life world. In other words, "Umwelt" refers to those features of a living being's surroundings that are meaningful to it.

Animals as Subjects

The function circle describes the Umwelt as a unity that is constituted by the world-as-sensed and the world of action. The function circle represents the organism as a subject that meaningfully integrates objects into its Umwelt. According to Thure von Uexküll, Uexküll's elder son, the function circle of his father can be seen as a semiotic activity in which the process of semiosis is manifest.[3] Indeed, as Jakob von Uexküll says, "one can speak of functional cycles as meaning cycles, whose task is determined to be the utilization of carriers of meaning."[4]

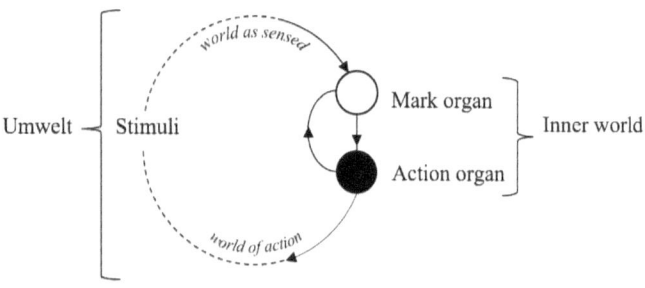

The function circle

Figure 1 The function circle

The indications of which the world-as-sensed consists are not mere copies of features of external entities. Rather they are constructed in a nontrivial cognitive process as spatially, temporally, and spatiotemporally localized features of the perceived world.

Many indications are combined into a *thing* (*Ding*).[5] A thing is a coherent unit of indications that occupies a moment and a place or a direction in space. It is an instantaneous datum of experience. Things are events rather than persistent entities. Animal and human subjects synthesize things unconsciously.[6] The unconscious creative process also creates another kind of cognitive entity—*objects* (*Objekte*). An object is an enduring thing, a thing extended in time. It is an enduring sequence of data of experience that occupies a particular spatiotemporal region in the subject's perceptual-space-time. Objects constitute higher units of experience than things.[7] Objects can be involved in lawful causal relations. Uexküll calls objects that possess a *framework* (*Gefüge*), merging their parts into organized whole *implements* (*Gegenstände*). Implements are objects in which "the parts stand in the same relation to the whole as the individual sounds to the melody."[8] Implements are organized wholes of data of experience. They might represent artificial or natural entities. The only natural implements that Uexküll knows are representations of organisms, parts of organisms (cells, tissues, organs), and groups of organisms.

Things, objects, and implements are not ontological but epistemological concepts. The world-as-sensed and the world of action of both humans and most animals is constituted by these three kinds of cognitive entities. According to Uexküll things, objects, and implements are differently complex products of one and the same unifying process, the so-called *apperception process*.[9] The apperception process lies at the root of all perception:

> Whatever the perception, the activity is of the same kind; different qualities are constantly being associated into unities. The power of the subject (*Gemüt*) that exercises this apperceptive activity is for ever creating new structures; in its very nature, it is a formative force (*Bildungskraft*).[10]

Uexküll's epistemologically founded biology of subjects is anchored in the assumption that the laws forming our attention and thus creating the Umwelt of our own subjectivity can also be recognized in animal subjects.[11] Uexküll makes clear that the apperception process, although lawful, cannot be mathematically described.[12] For this and other reasons biology cannot be reduced to physics.[13]

Uexküll's conviction about the non-reducibility of biology to physics is supported by Kant's concept of *pure* or *original apperception*, which is the underpinning philosophy of the Uexküllian apperception process. In his *Critique of Pure Reason* (*CPR*) Kant introduces pure apperception as a spontaneous a priori activity of the *subject* that synthesizes the manifold of our representations to a unity.[14] Pure apperception is an operation of what Kant describes as 'understanding' (*Verstand*):

> Combination does not lie in the objects, however, and cannot as it were be borrowed from them through perception and by that

means first taken up into the understanding, but is rather only an operation of the understanding, which is itself nothing further than the faculty of combining a priori and bringing the manifold of given representations under unity of apperception, which principle is the supreme one in the whole of human cognition.[15]

The unity of perceived data in all our representations "can be executed only by the subject itself,"[16] i.e., by a transcendental factor that can never be an empirical content of human representations. Unification of representation requires unity of consciousness. The logical consequence is:

[T]he unity of consciousness is that which alone constitutes the relation of representations to an object, thus their objective validity, and consequently is that which makes them into cognitions.[17]

Since Kant's term "understanding" refers to the *a priori* faculty of reason to combine representations and to transform their manifold into a coherent unity of apperception, it is a technical term and must not be confused with the common meaning of that term in everyday language.

Although Kant intended his theory of pure reason to provide a theoretical grounding for Newtonian physics, his conviction that the unity of experience is executed only by the subject goes against the objectivism and anti-transcendentalism which characterizes contemporary physics and the physics of Uexküll's time. From the materialist perspective, only a human or animal brain can cause the unity of human or animal experience. Both Kant and Uexküll would reject this contention, because according to both Kant and Uexküll brains are not transcendental factors but physical entities that can be sensually experienced in cognitive acts. Brains are not *a priori* principles but *a posteriori* data of perceptions.

Uexküll extended Kant's theory of human subjectivity to a general biological theory that applies to both human and animal subjects. He considers human and animal subjects to be transcendental; they are considered spatially nonlocalizable unities of apperception. The apperception process unfolds lawfully and determines the synthetic process of perception. *For that reason the apperception process can be considered the central category of subjectivity.* In this sense it is comparable to the striving of "actual occasions," as Whitehead describes the most elementary entities of reality in his metaphysics, to complete themselves as subjects.[18] Both Whiteheadian actual occasions and Uexküllian apperception processes are synthetic activities or, more aptly, *agents of concretion.*[19]

Continuity and Discontinuity between Human and Animal Spontaneity of Cognition

Uexküll's intellectual closeness to Kant suggests that he implicitly thinks of the apperception process as a spontaneous activity. For Kant, spontaneity is, in contrast to receptivity of (sensual) intuition (*Anschauung*), a cognitive faculty only of understanding (*Verstand*) and reason (*Vernunft*).[20] Since the meaning of "intuition" in Kant's works refers to sensual perception, it must not be confused with the meaning of this term in contemporary everyday language or in this chapter. The autonomous ability of the subject to form concepts makes thinking possible. Spontaneity and receptivity complement each other and produce cognition or knowledge, which arises only from the connection of (sensual) intuitions and thinking. Kant's famous quote "thoughts without content are empty, intuitions without concepts are blind"[21] perfectly characterizes the synergy of spontaneity and receptivity. It characterizes every human cognitive act as a spontaneous synthetic activity in which particular sensual representations become merged with general concepts.

From a Kantian perspective the most essential feature of human subjectivity consists in a specific connection of spontaneity and receptivity that is the very essence of pure apperception:

> The I think must be able to accompany all my representations (*Vorstellungen*); for otherwise something would be represented in me that could not be thought at all, which is as much as to say that the representation (*Vorstellung*) would either be impossible or else at least would be nothing for me. That representation that can be given prior to all thinking is called intuition (*Anschauung*). Thus all manifold of intuition has a necessary relation to the I think in the same subject in which this manifold is to be encountered. But this representation is an act of spontaneity, i.e., it cannot be regarded as belonging to sensibility (*Sinnlichkeit*). I call it the pure apperception, in order to distinguish it from the empirical one, or also the original apperception, since it is that self-consciousness which, because it produces the representation I think, which must be able to accompany all others and which in all consciousness is one and the same, cannot be accompanied by any further representation.[22]

The spontaneous activity of pure apperception is a self-determined factor of cognition. As such it makes human freedom possible. According to Kant, freedom is a faculty of practical reason. This means that it can be assigned only to subjects which are not only able to make *moral judgments* about real or hypothetic situations but, above all, to conceive those judgments as their own thoughts: the "I think" must accompany all their representations and thoughts. This ability, however, can hardly be ascribed to animal nonhuman species. The "I think" is not a sensual representation but itself a thought that, as with all thoughts, can only be produced by rational beings that operate with concepts, i.e., by beings endowed with symbolic capability. Thus, from a strict Kantian perspective it is difficult to assign freedom to animal subjects that lack the ability to operate with concepts and thus to produce thoughts, at least in our human understanding of thinking, not to speak of conceiving them as their own products. Uexküll's biology cannot account for creativity, which, in the strict meaning of that term, applies to *human* freedom. Nevertheless, this does not exclude other forms of freedom, beside human freedom. Kant's idea "thoughts without content are empty, intuitions without concepts are blind" must undergo a radical reinterpretation. That spontaneity and receptivity complement each other and produce cognition may also be seen to apply to animals.

It seems clear that Uexküll's theory of meaning displays latent semiotic aspects in that organisms operate and interpret *signs* that are related to entities or events of their Umwelts. In an essay written by myself and Arthur Araujo, drawing on Uexküll's *Theoretical Biology* (1926) and *A Foray into the Worlds of Animals and Humans* (1934), we suggest that *organisms are interpreters of signs*. Ascribing the ability of interpretation to animals requires the introduction of an extended concept of spontaneity that is not limited to the rational usage of human concepts and symbols. We suggest that Uexküll's perspective is compatible with a general biological form of transcendental spontaneity, a form of spontaneity more fundamental than Kant's pure apperception, which applies only to human consciousness.[23] According Uexküll, spontaneous processes merge a manifold of signs (signals) that are correlated with particular entities of the animal's Umwelt to form units of meaning. Those units are acts of experience, i.e., they are qualitative phenomena and not mere epiphenomena—they are complexes of *qualia*. In other words, animals, especially the most primitive ones, do not reflect rationally about the meaning of their perceptions. Their behavior is responsive to feeling. As Whitehead would say, apart from (animal) feelings there is no interpretation in the (animal) world. If this is true, *biological spontaneity synthesizes signs (signals) to units of feeling that have their own unique complex mental structure of qualia.* Such a structure can be seen as the *mental pole* of a Whiteheadian actual occasion.[24] Most importantly, Whitehead's concept of *concrescence* opens a new way of conceiving the process of amalgamation of signs to a complex feeling beyond antiquated substance ontology and physicalistic metaphysics. In contemporary philosophy of mind literature it is a commonplace understanding that qualia cannot be reduced to physical states. Hence, it is not possible to explain the synthetic process that generates those qualia by referring to mere physical factors, i.e., to spatiotemporally localized entities or events, such as brain processes. Consequently, not only human but also animal interpretation cannot be reduced to processes that are determined by the physical interactions between the sense organs of the organism and its physical surroundings (Uexküll's *Umgebung*). Following Kant, Uexküll claims that the apperception process generates the unity of human ego.[25] I suggest that the more basal biological apperception process that creates the unity of experience as a coherent qualitative phenomenon can be attributed even to primitive animal experiences. Thus, even if the Uexküllian animal subject lacks the ability to create and use human concepts, animals can still be understood as sources of *creative* spontaneous processes that display a *rudimentary form of freedom*. The basal biological apperception process that synthesizes the unity of any process of experience must be creative because—as being a specific form of the Kantian apperception activity—it cannot be determined by any physical factors, i.e., by sensually experienceable entities. Thus it must occur spontaneously and manifest an elementary form of freedom that could be described as "proto-freedom."

There are similarities and dissimilarities between animal and human forms of spontaneous cognition. As human beings, we have the unique ability to experience and express our selves through language. Humans are the only animals able to create abstract concepts and put them to use in our daily lives. Kant's fundamental insight "The *I think* must be able to accompany all my representations" still expresses the most essential truth about the human mind, because the "I think" mostly unconsciously penetrates all our cognitive acts and forms them. In addition, this insight applies only to human subjectivity because it is true only for beings able to operate with linguistic terms. We are the only known biological beings that do not only have a self but also an abstract ego. *Only human subjects can develop abstract egoity*. This doesn't, however, mean that the human self is restricted to abstract ego. We share with animals the spontaneous, creative, and deeply unconscious activity of merging a manifold of signs into units of meaning. The nonverbal cognitive activity in adults (dreams, altered states of consciousness, spontaneous insights, etc.) and young children without faculty of speech proves that those processes are not primarily determined by language skills and abstract concepts.

CASSIRER'S UNDERSTANDING OF THE HUMAN AS ANIMAL SYMBOLICUM AND BERGSON'S CRITIQUE ON HUMAN INTELLIGENCE

In his book *An Essay on Man*, published in 1944, Ernst Cassirer (1874–1945) considers Uexküll to be "a defender of the principle of the autonomy of life": "Life is an ultimate and self-dependent reality. It cannot be described or explained in terms of physics or chemistry."[26] Uexküll's primarily epistemological approach to biology, according to which animals and humans are subjects that build "a world in itself," emphasizes the *mediatedness of cognition*.[27] This idea, together with the introduction of the apperception process as a synthetic activity governed by *a priori* forms, had to attract Cassirer's attention since he was strongly influenced by the mathematically-scientifically oriented Neo-Kantian "Marburg School."

However, with respect to human cognition, Cassirer expands Uexküll's function-circle by a component, "which appears to be the distinctive mark of human life."[28] In humans, between the "world-as-sensed" and the "world of action" we find a "third link," which is the world of symbols.[29] The human lives "in a new *dimension* of reality," in a "symbolic universe," parts of which are language, myth, art, and religion.[30] The human "cannot see or know anything except by the interposition of this artificial medium."[31] Thus "instead of defining man as *animal rationale* we should define him as an *animal symbolicum*."[32]

Nonhuman animals understand and use signs. Nevertheless, Cassirer makes a distinction between signs and symbols. Animals with highly developed nervous systems are able to express emotions, such as rage, terror, desire, playfulness, and pleasure by means of gesture. But *animal communication lacks symbols, which are signs with an objective reference or meaning*.[33] According to Cassirer "(t)he difference between *propositional language* and *emotional language* is the real landmark between the human and the animal world."[34]

This difference, which makes humans the only "symbolic species"[35] of Earth, characterizes also the specific difference between human and animal intelligence. As Cassirer argues, animals possess "a practical imagination and intelligence, whereas man alone has developed a new form: a symbolic imagination and intelligence."[36]

Cassirer highlights three crucial differences between human language and animal usage of signs. *First*, he illustrates that symbolization is "a principle of *universal* applicability," since everything can be denoted.[37] Other than signs used or interpreted by animals, which represent specific entities, situations, or emotions, symbols are not restricted to particular cases. *Secondly*, Cassirer explains how a symbol is "extremely variable [. . .] (whereas) a sign or signal is related to the thing to which it refers in a fixed and unique way."[38] Whereas any one individual sign or signal refers to a certain individual entity or process, a specific idea or thought may be expressed by using quite different symbols or combinations of symbols. *Thirdly*, Cassirer identifies human language as able "to isolate relations—to consider them in their abstract meaning."[39] By using an adequate symbolism humans are able to abstract from particular entities and to study their spatial and other relations to a degree that is far beyond animal cognitive faculty. Geometry and algebra are the classical examples of the human's ability to study universal relations in abstraction from related entities. Without the preliminary step of human language, mathematics would not be possible.

Biologic Relevance of Symbolic Systems: a Central Cause of Discontinuity between Human and Animal Evolution

Language is a main causal factor that has influenced the evolution of the human brain.[40] In addition to brain evolution, linguistic development directed physical activities in specific ways and changed humans' dietary habits. This has decisively influenced the evolution of our whole body; for example, hands, teeth, muscles, and our digestive system. The form of our skeleton and the structure of our muscles, bones, sinews, and nerves has evolved to such an unprecedented degree that human beings' dexterity, coordination, and complex motion abilities far exceed those of other animals. Just as no animal can write, there are no animals able to perform a ballet dance or synchronized swimming, play piano, or free-climb.

Beside natural language, the abstract languages of mathematics, philosophy, sciences and technology have exerted an enormous impact on human evolution. One need only think of how the Neolithic technique of stonecutting and ceramics, the invention of the plow, and the emergence of metal have formed agriculture and urban life, which themselves have critically changed the route of our physical evolution. Of course, these new techniques and technologies require the invention of a specific terminology and thus the evolution of language. Through the scientific-technological revolution of the last centuries, abstract and formal symbolic systems have become exceedingly complex and have influenced our ongoing evolution. Since the *Weltanschauung* of that revolution has been influenced by Renaissance Platonism, and later by materialistic, dualistic, and empiristic philosophies, those philosophic traditions have had and continue to have a crucial effect on our evolution. Scholars of process philosophy, including the authors featured in this volume, hope to undermine the dualistic scientistic-technocratic logic that has been so detrimental to planetary evolution and to offer new ways for understanding the natural world and our place within it, ways that contribute to rather than diminish the health of the planet.

Since Darwin's introduction of the concept of *sexual selection* in his work *The Descent of Man, and Selection in Relation to Sex* (1871) we know that the sense for beauty and the sexual attraction caused by it are powerful factors of both animal and human evolution.[41] Our symbolic systems have guided our sexual selection in such a highly subtle way that it has scarcely been adequately recognized yet. For thousands of years the interpenetration of specific symbolic systems, such as literature, myths, and legends with music, dance, and forms of body beautification has played a crucial role in social events, such as celebrations and athletic contests in honor of gods, in which potential partners have met one another. In many cultures skills shown in those cultural events have been, and still are, an important factor for marriage. Due to the sexual revolution and liberation of the last century this tendency has been considerably intensified: one needs only to think of the millions of partnerships that began in dancing sessions, parties, and rock festivals in the last fifty years. If we consider that showing up to those events requires participation in complex networks of symbolic systems, the role of symbolic forms in erotic attraction becomes obvious.

Human Abstract Perception of Space and Time and Bergson's Concept of Spatialization

A direct result of the ability of human species to use geometrical and other mathematical symbolisms to focus on abstract spatial relation is the abstract perception of space. Unlike animals, which live in their individual concrete "perceptual space," humans are able to conceive the idea of "abstract space" or "symbolic space" by a very complex process of thought.[42] Since the time of Newton, physics has been based around abstract or mathematical space, which should not be confounded with the space of our sensual experience. Abstract space is conceived of as an entirely homogeneous extension that is a fiction of the human mind; it does not represent any physical or psychological reality. Cassirer considers the "points and lines of the geometer [. . .] [to be] nothing but symbols for abstract relation."[43]

Cassirer was familiar with the works of Henri Bergson (1859–1941), a process philosopher. In his first book, *Time and Free Will* (1889), Bergson reflects on the fact that within the field of mechanistic physics mathematical space is understood as an empty vessel, the parts of which are considered many simultaneously existing but distinctive and smaller empty vessels:

> [It is the] clear idea of a homogeneous medium, i.e., of a simultaneity of terms which, although identical in quality are yet distinct from one another.[44]

He thinks that this conception of space in mathematical physics has very deep roots and can be traced back to the beginnings of arithmetical thinking. According Bergson, there is an intrinsic connection between geometry and arithmetic. Bergson's point of departure is that it is not possible to conceive the idea of addition in a universe in which there is time but no space:

> It is certainly possible to perceive in time, and in time only a succession which is nothing but a succession, *but not an addition*, i.e., a succession which culminates in a sum. For though we reach a sum by taking into account a succession of different terms, yet it is necessary that each of these terms should remain when we pass to the following, and should wait, so to speak, to be added to the others how could it wait, if it were nothing but an instant of duration? And where could it wait if we did not localize it in space? We involuntarily fix at a point in space each of the moments which we count, and it is only on this condition that the abstract units come to form a sum.[45]

Entities to be added are "parts of space, and space is, accordingly, the material with which the mind builds up number, the medium in which the mind places it."[46] Simultaneous coexistence of entities to be added is, however, not the only necessary condition for addition. Only entities considered to have one and the same essence can be added.[47] This condition is best fulfilled by entirely abstract entities, the essence of which is pure quantity bare from any quality. Clearly, numbers are those entities:

> [T]he idea of number implies the simple intuition of a multiplicity of parts or units, which are absolutely alike.[48]

A Newtonian understanding considers empty space as pure extension bare of any possible quality and thus as constituting an entirely homogenous continuum. Of course, this applies equally to all parts of empty space, i.e., to all its possible divisions into smaller empty spaces. Therefore *empty space and numbers are in some sense coessential.* There is an "inter-connexion between the notions of number and space."[49] Since Bergson considers numbers, like all concepts, creations of the human mind, he claims that the abstract idea of homogeneous space grows out of an effort of human intelligence and concludes:

> [T]he higher we rise in the scale of intelligent beings, the more clearly do we meet with the independent idea of a homogeneous space. It is therefore doubtful whether animals perceive the external world quite as we do, and especially whether they represent externality in the same way as ourselves. [...] This amounts to saying that space is not so homogeneous for the animal as for us, and that determinations of space or directions do not assume for it a purely geometrical form. Each of these directions might appear to it with its own shade, its peculiar quality.[50]

From a Bergsonian perspective, the transition from animal experience of a spatially extended qualitative manifold to the purely quantitative, geometrical space characteristic of the most abstract forms of human experience constitutes a major discontinuity in the evolution of perception. This transition was driven by the human invention of specific symbolic languages that allow the mind to grasp and operate upon numerical, geometrical, and mechanical concepts.

Beside the experience of space, human symbolisms radically influence our experience of time as well. "When dealing with the problem of organic life," says Cassirer, "we have, first and foremost, to free ourselves from what Whitehead has called the prejudice of 'simple location.' The organism is never located in a single instant."[51] The momentary state of an organism cannot be described without taking that organism's history into consideration and without referring to its future. Cassirer understands memory to be a general function of all living beings, meaning that the organism preserves in its body material traces of past events and that these traces influence its future reactions. He makes clear, however, that human memory is something quite different. Other than in animals, human recollection cannot be described as an ideational return of past events as a faint copy of former experiences. It is rather "a rebirth of the past; it implies a creative and constructive process."[52] Human memory is a *symbolic memory*, which is "the process by which man not only repeats his past experience but also reconstructs this experience. Imagination becomes a necessary element of true recollection."[53] In his second book, *Matter and Memory* (1896), Bergson emphasizes the creativity of the process of recollection as well.[54]

In *Time and Free Will* the most central issue Bergson deals with is human experience of time through the lens of the conception of abstract space or, in other words, the *spatialization of time*.[55] Modern scientific thought, which has been influenced by physics, has a strong tendency to treat time as an infinite homogeneous medium. This reduction of time to pure and objective quantity bare of any experiential qualities is not limited to several subfields of physics but has been expanded also to the study of biological and psychological processes. It has also occupied the perception of time in everyday life as well as most humans' understanding and even experience of the temporality of their own mental and emotional life. Nevertheless, if psychical time were nothing but a sort of one-dimensional, totally homogeneous medium, we would have to consider our mental processes as successive events that occupy distinct positions in this medium, just as physical bodies in empty space do. In other words, the processes of our inner life would be sharply separated from each other and thus not interpenetrate one another at all. As Bergson, however, correctly claims:

> States of consciousness, even when successive, permeate one another, and in the simplest of them the whole soul can be reflected.[56]

This follows directly from Bergson's concept of *duration* that is not only the main pillar of his philosophy of time but also builds the basis of his metaphysics. The most essential feature of human consciousness is that our mental states penetrate one another so that their interconnection organizes them in such a way that each one of them reflects the whole temporal continuum of our experience:

> Pure duration is the form which the succession of our conscious states assumes when our ego lets itself live, when it refrains from separating its present state from its former states. [...] it is enough that, in recalling these states, *it does not set them alongside its actual state as one point alongside another*, but forms both the past and the present states into an organic whole, as happens when we recall the notes of a tune, melting so to speak, into one another. Might it not be said that, even if these notes succeed one another, yet we perceive them *in one another*, and that their totality may be compared to a living being whose parts, although distinct, *permeate one another* just because they are so closely connected? [...] We can thus conceive of succession without distinction, and think of it as a *mutual penetration*, an interconnexion and organization of elements, each one of which represents the whole, and cannot be distinguished or isolated from it except by abstract thought.[57]

Accordingly, the contents of our mental life are not distinct from one another like bodies in empty space that are clearly separated and therefore *countable*. Thus the idea of countability, so fundamental to modern sciences, is not possible within human temporality or duration. The duration of our experience constitutes a *heterogeneous continuum*: the stream of interconnected experiential qualities that "permeate one another." These mental acts determine their own essence through their interpenetration. *The whole duration is a process that permanently transforms its own essence.* Thus, it is in the most radical sense of the word a *heterogeneous* multiplicity. In sharp contrast to the homogeneous continua constructed by abstract intelligence, heterogeneous continua constitute the most concrete continuum that we know. Because heterogeneous continua incessantly transform their own essence, they cannot be subject to any mathematical operation. Since the mental acts are internally related one to another, they cannot be separated from each other, so that duration can neither be increased by addition of distinct mental acts, nor divided by those acts. In other words: there are no "pieces of duration." Thus, due to its radical heterogeneity, duration might be interrupted by an external cause but not divided by it. Duration is a creative continuum that continues itself.

Bergson assumes that the abstract considerations of space and time as homogeneous empty media are not equally fundamental and primordial and "that the idea of space is the fundamental datum."[58] The abstract idea of time as a homogeneous and unbounded medium has been derived from abstract space:

> Time, conceived under the form of a homogeneous medium, is some spurious concept, due to the trespassing of the idea of space upon the field of pure consciousness.[59]

Therefore, Bergson makes a sharp distinction between two different kinds of reality: heterogeneous reality and homogeneous reality. The former is the reality of sensible qualities. The latter includes abstract space, scientific conceptions of time, mathematics, and other abstract constructions.[60]

> This latter [the homogeneous kind of reality], clearly conceived by the human intellect, enables us to use clean cut distinctions, to count, to abstract, and perhaps also to speak.[61]

Bergson's conception of "spatial" order has been widely understood in terms of *spatialization* in the field of philosophy. The concept of spatialization not only implies that "every homogeneous and unbounded medium"[62] is spatial, but, even more, it allows us to reassess our scientific, economic, and other abstract theories through Bergson's above-discussed insights about the essence of spatial and homogeneous order. It is especially important in this context to consider that "if we notice that abstraction assumes clear-cut distinctions and a kind of externality of the concepts or their symbols with regard to one another," we shall understand that "the faculty of abstraction" in general is implicitly based on the abstract conception of a "homogeneous medium."[63] In other words, *all* our abstract concepts, symbols, and theories are to some degree specific manifestations of what Bergson calls "spatialization."

From a Bergsonian perspective, the human faculty of spatialization lies at the root of the discontinuity between human and animal life. As it is intrinsically and indissolubly connected with intelligence, spatialization can be seen as a main cause of the arising of human nature and the bifurcation of human species' evolution from the evolution of all other primates. As already noted, Bergson argues that "the higher we rise in the scale of *intelligent* beings, the more clearly do we meet with the independent idea of a homogeneous space."[64] Bergson's theory of the evolutionary relationship between intelligence and homogeneous space is unpacked in his best known book, *Creative Evolution*. In the second chapter of that book he makes his famous distinction between *instinct, intelligence,* and *intuition*.[65] As René Pikarski makes clear in the present book, whereas instinct and intuition are faculties of mental closeness between subject and object, intelligence is the capability for cognitive and emotional distance between them. This is due to the indirect, mediated contact, by more or less abstract thoughts, of intelligent beings with nature. According to Bergson, intelligence is a mental faculty of not only humans but also of other living beings. Intelligence is an essential feature of vertebrates as well.[66] Humans, however, are the only species that has developed this ability to unprecedented heights. Human experience of space, time, matter, and other living beings has been shaped by our spatialized intelligence—first and foremost by our abstract scientific symbolisms that homogenize the intrinsic heterogeneity of all expressions of life.

At this point, it may be helpful to point out that in this chapter the terms "intelligence" and "intellect" have the same (Bergsonian) meaning that Alex Gomez-Marin and René Pikarski imply when they use these terms elsewhere in this volume. This is not the case, however, for Theo Badashi's chapter. In Badashi's references to "Nature's Intelligence," the "deep intelligence of the universe," and "Cosmic Intelligence," he ascribes to "intelligence" a radically different meaning that in many respects coincides with the meaning of "intuition" in the present chapter.

The increasing dissociation of human symbolisms from living nature that lies at the basis of scientific and technological rationality and that is a central topic of Bergson's writings, seems to escape Cassirer's attention. But, on the other hand, he emphasizes an aspect of time that is overlooked in Bergson's writings: For an appropriate understanding of human relation to time the dimension of the future is even more crucial than the dimension of the past. Anticipation of future events and even preparation of future actions is an important factor in the life of animals with highly developed nervous systems. In humans, however, as Cassirer says, "(t)he future is not only an image; it becomes an 'ideal.'" Only humans are able to conceive of an idea of the future. Our symbolic forms enable us not only to expect the future but to upgrade it to an "*imperative* of human life." Cassirer calls our *symbolic future* a "prophetic future" because it is best expressed in the life of the great religious prophets.[67] These religious teachers did not simply foresee future events or warn of future evils. *Their prophecies were the exact opposite of auguries*:

> The future of which they spoke was not an empirical fact but an ethical and religious task. [...] Prophecy does not simply mean foretelling; it means a promise. [...] Here too man's symbolic power ventures beyond all the limits of his finite existence. But

this negation implies a new and great act of integration; it marks a decisive phase in man's ethical and religious life.[68]

From a Bergsonian perspective, one could be tempted to say that there is an *essential* difference between, on the one hand, the theoretical or abstract idea of the future in which all adult and healthy humans are able to participate and, on the other, the prophetic future. A distinction that one frequently, and in respect to totally different issues, encounters in Bergson's books is that between "difference of nature" and "difference of degree." Whereas the theoretical idea of the future requires common human intelligence, only humans that have developed their intuition far beyond the level of the average human's intuitive abilities may be open to the prophetic future. The difference between both forms of foreseeing the future corresponds with the difference of nature between intelligence and intuition, as we will see later.

HUMAN UMWELT—AN ETHICAL IMPERATIVE

The human Umwelt does not merely have a threefold structure—"world-as-sensed," "world of symbols," and "world of action"—but rather these three dimensions indissolubly interpenetrate one another. Kant's famous slogan "intuitions without concepts are blind" anticipates Cassirer's insight that we cannot even see anything except by the interposition of symbols. Our conceptual denotation of objects essentially influences our perceptual experience of them. To be an animal symbolicum means to perceive the world through abstract "organs" formed by millennia-old cultures. This symbolic mediatedness necessarily increases the distance of human intellect from what Uexküll calls the "world-as-sensed" and the "world of action." Our highly entangled symbolic forms not only allow for understanding our world, *they also restrict our comprehension of what we perceive and how we affect our Umwelt.* Paradoxically, this distance from our Umwelt has made possible for us the extreme extension of both our world-as-sensed and our world of action through the aid of artificial devices, such as telescopes, microscopes, and particle accelerators. The development of these material devices is based on our most powerful, because most universal, instruments—our scientific concepts[69]—which can be invented only within advanced symbolic systems.

The vast variance of our world of symbols extends the human Umwelt far beyond the Umwelts of animal species, which are limited by their sensual perceptions. Each theoretical and technical discipline, all arts, and all political discourses constitute a meaningful world and hence an Umwelt. Thus we all live and act in many intersecting symbolic Umwelts, each of which is inhabited by a huge number of abstract concepts. One of the most important symbolic worlds is our *ethical Umwelt*.

Our symbolic Umwelts of physics, chemistry, and biology have made the infinity of space and time objects of our scientific research. They have made it possible for us to think systematically about the vastness of space, the past and the origin of the universe, and the evolution and origins of life. However, our abstract, purely symbolic access to these areas of physical actuality does not guarantee that we *understand* the symbolized entities and processes. It was not by chance that German Neo-Kantian philosopher Heinrich Rickert introduced a distinction between *understanding* (*Verstehen*)—a concept, however, that must not be equated with Kant's concept of understanding (*Verstand*)—and *explaining* (*Erklären*). Experience of value and meaning is the *conditio sine qua non* for understanding. Hence, entities and processes that do not have any value or significance for us cannot be understood but merely described or explained.[70] From the perspective of Edmund Husserl's phenomenology, we may say that we understand only beings and processes, which are part of our *life-world* (*Lebenswelt*) or realm of our sensual and other experiences.[71] From Rickert's and Husserl's point of view we *cannot* say that we understand physical and biological entities and processes, which we can explain by applying our abstract symbolisms, if they do not belong to our life-world. "Understanding" the explanations of scientists is not the same as understanding the beings the explanations are about. "Understanding" abstract symbolic systems, such as mathematical and logical equations and algorithms, is possible only because of internal explanation. Or stated another way, we acquire an understanding of such algorithms and equations by actually explaining them to our selves.

Uexküll divides our visual area into the "visual space" and the "remotest plane."[72] Within the visual space we are able to see objects stereoscopically and thus to have depth perception of them. In other words, we only perceive our spatial distance from objects if they are within our visual space. The outer limit of our visual space is the remotest plane. If objects are beyond our remotest plane we are not able to estimate which of them is closer to us and which is further. We perceive such objects as though they were placed on the inner side of the same spherical surface, the so-called "celestial sphere." All celestial bodies appear to move on that sphere. In direct analogy to Uexküll's distinction between visual space and space beyond our remotest plane, we may also separate our symbolic processes into those inhabiting our "area of understanding" and those operating beyond our "remotest plane of understanding," that which marks the beginning of a vast "space" of knowledge and that may be called "space of mere explaining." Of course, the usage of the term "space" here corresponds well with Bergson's understanding of spatialization or spatialized (abstract) knowledge. We can only then be confident that we understand beings and processes if they inhabit our "area of understanding," our life world. With regard to all of the other phenomena outside our life world we can only explain them. Viewed in this light we should not think that we understand the essence of entities, which we denote by scientific symbols, such as "electrons," "quarks," "quantum processes," "gravitational waves," "dark energy," "black holes," "genes," and "proteins" without them being a part of our embodied, experiential world or life-world (Husserl), as are trees, humans, rocks, mountains, oceans, storms, feelings, thoughts, and many of our own organic processes.

The entirely abstract concepts of contemporary physics, life sciences, technology, and biotechnology are clearly outside of what I have called our "area of understanding." The fact that we successfully operate with abstract symbols in our scientific languages proves only that we have learned the abstract rules of their application; it by no means shows that we understand the nature of the represented entities, let alone the complex relations between them. The symbolic systems of contemporary nano- and biotechnology, to most inventors of which the concept of life-world doesn't mean anything, allows manipulating natural beings without having even the faintest idea of the tremendous distance between their nature and our explanations of them, since those entirely abstract concepts are clearly outside of what I have called the "remotest plane of understanding."

Unfortunately, this negative aspect of symbolization, which beside science and technology also haunts politics, Anglo-American analytic philosophy, and neoliberal economics, seems to escape Cassirer's attention in *An Essay on Man*. Of course, in 1944 it was not nearly so obvious as it is today that our ignorance of the distance between our abstract symbolisms and the nature of the symbolized entities and processes—the basis of what Whitehead so accurately described as "fallacy of misplaced concreteness"[73]—can be so destructive. Today we have to understand that explaining should not be confused with understanding and that the horizon of our life-world grows incomparably slower than our ability to act outside of our "area of understanding." Moreover, my impression is that the horizon of our life-world is collapsing at ever increasing speed due to the constantly accelerating spread of all forms of digital "communication" and other technologies, especially those that promise to control the future of our life as well as that of other species, for example, biotechnologies.

Nowadays it is obvious that, as Matthew T. Segall says, "symbolic consciousness also has the power to produce civilizational myths that are entirely detached from the ecological context of the living planet that sustains us" (Chapter Two).

SACRED ENVIRONMENT AND THE RISING OF THE AGE OF INTUITION—A BERGSONIAN-WHITEHEADIAN PERSPECTIVE

The term "Umwelt" cannot have the same meaning that Uexküll gave to it over a hundred years ago. Today, "Umwelt"' can no longer simply mean the part of our surroundings that is meaningful to us. In today's German language "Umwelt" means "environment." However, in different discourses "environment" has different meanings. From the scientific point of view of theoretical ecology, both the rainforests of Earth and the dunes of Mars are environments. But *what is at stake today is the rescue and preservation of the living Umwelt of the Earth* and not the "terraforming" of Mars that is promoted by some contemporary technocrats. Given the current severe ecological crisis, it is imperative that any understanding of "Umwelt" has an ethical imperative. I suggest the following definition of "Umwelt" or "environment," to wit: "the *living* world to be saved." This world has a spatiotemporal extension. Its spatial extension coincides with the terrestrial biosphere. Its temporal extension entails the past and, most notably, the future of the biosphere, which includes the future of humanity. Thus the term "Umwelt" must refer also (but not only) to *future* living beings including humans. We should, however, not forget that the survival of our biosphere will be decided in the next decades and not in a distant future. The term "the living world to be saved" is an intrinsically political concept, laden with strong ethical intentions. *This term refers to something that must be saved and preserved because it is indispensable and, at the same time, it is in severe danger.* From this point of view the concept of Umwelt/environment should not be applied to other planets or space colonies. In our extremely critical present age it is important to outline the concept of "Umwelt" as an *Earth-centered* or *geocentric concept* because what is at stake is the rescue of *this* world, in which we live, *now*. We have to get rid of the technocratic temptation to think of possible "terraformed biospheres" on other planets as if this would be just a matter of scientific knowledge, economic power, and time.

The concept of Umwelt/environment, as I understand it, has to be reinterpreted in the light of both Cassirer's pioneering concept of the "prophetic future" and Bergson's critique of spatialization. As stated previously, prophecies are not about future events but about *promises, the fulfilling of which is an ethical task*. This, however, I quote again, "implies a new and great act of integration." What else should this integration be today than the integration of science, technology, economy, ecology, and ethics? Cassirer's concept of *prophetic future* motivates us to consider how to reconcile the competing interests of science, industry, and the financial sector with what I have called above an "ethical Umwelt."

This integration requires a view of nature formed by a *new mental closeness or intimacy of understanding* as a counterbalance to the emotional distance of scientific explanations and technological applications. From the perspective of Rickert and Husserl *understanding evolves out of the experience of value*. This raises, of course, the question "for whom do organisms have value?" As Kant says in the *Critique of Judgment*, an organism "can be called a *natural purpose*, and this because it is an *organised* and *self-organising being*."[74] As being purposes for themselves, organisms may also be considered self-values. In the twentieth century, the influential philosopher Hans Jonas based his metaphysics of the organism on the concept of *freedom*.[75] This is in accordance with Kant's seminal insight that organisms act and exist for their own purpose. Thus, following Kant and Jonas, in order to understand living nature and not just to explain it, we must be able to experience living beings as manifestations of intrinsic values, which means that they should not be valued for the sake of their contribution to some ends desired by humans, but for their own sake. This is true from Whitehead's perspective as well. As Segall makes clear "[t]o be actual, to be a fact, for Whitehead, means to experientially enjoy existence as an end in itself, to value oneself as an actuality and to be valued by other actualities" (Chapter Two). Thus, the term "ethical Umwelt" refers to all living beings of the present and the future as being intrinsic values.

Whitehead's metaphysics provides an excellent philosophical foundation for ethical Umwelt.[76] Our current mainstream scientific worldview supports the reduction of living beings and natural processes to passive valueless entities devoid of any kind of striving and feeling. The reduction of living beings to passive entities is, for example, an assumption of neoclassical economics, which understands economy as something that functions in isolation from ecological, social, end ethical issues.[77] The economy operates based on the abstraction of different forms of capital. For example, natural resources, financial capital, human capital, and know-how are considered quantifiable and therefore convertible and thus interchangeable.[78] This strategy serves the interests of a particular "moral community" that consists of contemporary mainstream Western economics, industry, and politics, i.e., of an "elite" of Western individuals living in the present. "Moral community" is here understood as the community with the power not only to protect its interests but also to force them upon others because its values are dominant and its knowledge is taken seriously by the majority of Western people. In diametrical opposition to this ideology, Whitehead, Bergson and other process philosophers provide avenues for developing a new economics based on the principles of non-convertability of different forms of capital,[79] an idea that perfectly corresponds with Bergson's appeal to do justice to heterogeneity. Process-philosophical economic and scientific theories embed economy and science in the biosphere. In this way such theories expand the moral community to include animals and plants, as well as present and future generations.[80] Whiteheadian metaphysics suggests that the value of future living beings is not a future value but a present value: *the future has its value now*. Process philosophy considers the extended moral community as being a part of an ethical Umwelt. The idea of a morally significant Umwelt implies its *sacredness*, i.e., that it is a *sacred environment*.

If the concept of ethical Umwelt must embrace all living beings of the future and the present as being intrinsically valuable, the question arises, how can this be attained? This, certainly, cannot be done ad hoc by an autonomous decision of human intelligence. The intrinsic value of nature, and especially of living beings, can only then become the main pillar of a new worldview if it is *empathically experienced*. This brings us back to Bergson's concept of *intuition*.

The word "intuition" stems from the Latin verb "*intueri*," which means "to look at." It denotes an immediate and nondiscursive form of cognition. *It must be noticed that Bergson uses that term in a radically different way than Kant does*. In *Creative Evolution*, Bergson considers life as "consciousness launched into matter"[81]—life is duration enveloped by matter. From this perspective, intuition is one of the two cognitive faculties in which the primordial cognitive activity of (ur)consciousness split up. *Intuition is the attention of consciousness on its own movement*.[82] (It is worth noticing that we are only then able to experience the nature of our duration as a radically heterogeneous continuum if we direct our attention to our own psychical processuality.)

> [A] glance at the evolution of living beings shows us that intuition could not go very far. On the side of intuition, consciousness found itself so restricted by its envelope that *intuition had to shrink into instinct*, that is, to embrace only the very small portion of life that interested it; and this it embraces only in the dark, touching it while hardly seeing it. On this side, the horizon was soon shut out.[83]

On the other branch of the bifurcation of consciousness, that of intelligence, consciousness succeeded in increasingly freeing itself from the restrictions that matter imposed on it:

> On the contrary, consciousness, in shaping itself into intelligence, that is to say in concentrating itself at first on matter, seems to externalise itself in relation to itself; but, just because it adapts itself thereby to objects from without, it succeeds in moving among them and in evading the barriers they oppose to it, thus opening to itself an unlimited field.[84]

In Bergson's view, the evolution of instinct reached its highest level in the insects and especially in the socially organized hymenoptera.[85] Intelligence emerged on the branch of evolution of the vertebrates in which consciousness focused its attention on the *manipulation of matter* it was passing through.[86] In other words, the whole evolution of the animal kingdom split up into two divergent branches: one which evolved instinct, the other intelligence.[87] In contrast to the highly specialized external organs of hymenoptera, such as the mouthparts and the sting of bees and the various mandibles of ants, the external organs of the vertebrates, especially the human hand, have not evolved in order to manipulate specific material structures. Instead of being adapted to the structure of particular material objects, human hands are the most generally usable organs of animal kingdom. The universalistic structure of the human mind, which consists in its unlimited ability to invent, use, and redefine abstract concepts, corresponds directly with the hand's universal abilities. Both human hands and symbolic forms are universal tools which enables us to manipulate an unlimited number of real and potential material structures. Due to the coevolution of the hand and the symbolic mind, the human species progressed much further than the social insects in freeing itself from the limitations of matter. However, "[o]nce freed, moreover, consciousness can turn inwards on itself, and awaken the potentialities of *intuition which still slumber within it.*"[88] The primordial interrelation of life and consciousness is not absent in modern humans—it only slumbers within us.

Since intelligence "goes all round life, taking from outside the greatest possible number of views of it, drawing it into itself instead of entering into it,"[89] "[w]e are at ease only in the discontinuous, in the immobile, in the dead."[90] Because of the modus operandi of intelligence, our "intellect is characterized by a natural *inability* to comprehend life."[91] From a Bergsonian perspective, it is the one-sided dominion of intelligence that has caused the enormous alienation of human science, technology, and economy from living nature, creating thus the disastrous contemporary ecological crisis. Before the background of the ethical imperative that obliges us towards what I have called "the living world to be saved," the following insight of Bergson deserves our whole attention:

> [I]t is to the very inwardness of life that intuition leads us,—*by intuition I mean instinct that has become disinterested, selfconscious, capable of reflecting upon its object and of enlarging it indefinitely.*[92]

> Instinct [...] is moulded on the very form of life. While intelligence treats everything mechanically, instinct proceeds, so to speak, organically. If the consciousness that slumbers in it should awake, if it were wound up into knowledge instead of being wound off into action, if we could ask and it could reply, it would give up to us the most intimate secrets of life.[93]

Intelligence is a cognitive faculty that operates with universal abstract symbolic systems which, because of their universality, cannot be adapted to the specific essence of the particular beings to which the intelligent creators of those abstract systems refer. Therefore, there is always an insurmountable cognitive distance between the intelligent subject and its object. Since "*instinct is sympathy,*"[94] that distance can only be bridged by the latter which is a cognitive ability that captures the nature of its objects through internal relations between subject and object. Thus, as René Pikarski makes clear in Chapter Four, instinct recognizes its objects empathically through a closeness of being and a sympathy towards life.

From the background of Whitehead's ontology, Bergson's understanding of instinct as sympathy (sym-pathy) between living beings can be best approached by the technical term "prehension," or, more precisely, "physical prehension."[95] The concept of prehension corresponds well with what in Western metaphysics has been described as "internal relations," i.e., as relations through which the related entities constitute themselves, thus being unable to exist without them. One of Whitehead's most ingenious insights that constitutes the very core of his ontology is that no prehending subject preexists its prehensions. Rather, it comes into existence by prehending other already constituted beings that become its objects. Accordingly, the subject emerges out of a cognitive act that is literally metaphysical (meta-physical), since it does not take place in the physical world. The subject arises as a spatiotemporally localized entity only after the integration of its prehensions into a coherent unity of experience. Of course, one could object that every bee is localized in space and time, and that it preexists its relation to any particular flower that it approaches. By "subject," however, Whitehead does not understand a living being, such as a bee or an ant, but an actual occasion, i.e., a short-lived process that determines its own constitution through its relations to the Umwelt (see note 18). As already noted, all actual occasions have a mental pole that in almost all cases does not go beyond a proto-mental level (see note 24 of this chapter). Nonetheless, actual occasions that occur in nervous systems, even in those of insects, are coherent complexes of experience. Their mental capacity exceeds by many orders of magnitude the niveau of proto-mentality. Thus, from a Whiteheadian perspective, although the bee is not a subject, its mental life is a series of short-lived subjects. Each subjective act through which the bee recognizes a particular flower emerges as a unification of prehensions that are empathic or sympathic relations. From that angle the subject is not a bearer of experiences but rather an act of cognition that arises together with its instinctive relations to the Umwelt and cannot be separated from them. It is this nonseparability of the processual subject from its object that constitutes the *closeness of being* or *nondiscursive immediacy* of instinctive cognition. Recently, similar positions have been adopted in the new field of *deep* or *radical neurophenomenology*.[96]

While instinct provides an immediate knowledge of living objects, it by no means imparts any understanding of them in the specific meaning that Rickert gave to this term. Understanding cannot be provided by a cognitive faculty that restricts itself to conveying a mere practical and thus limited knowledge of its objects. While real understanding demands that the subject embraces the essence of its objects and thus experiences them as intrinsic values and purposes for their own sake, instinct captures only those aspects of its objects that are useful to the subject. Only "instinct that has become disinterested, self-conscious," because the consciousness of that which that slumbered in it has awakened, is, as already said, "capable of reflecting upon its object and of enlarging it indefinitely,"[97] so that the subject can embrace the essence and intrinsic value of its objects. Instinct that is "wound up into knowledge instead of being wound off into action" has become intuition. As Alex Gomez-Marin so aptly states in Chapter Four, "[i]nstinct and intelligence are bound to survival, whereas intuition is self-transcendent in the sense that it becomes an understanding not restricted by (nor opposed to, perhaps just indifferent to) utility." In other words: *Real understanding, especially of living beings, can only be acquired by intuition*. Because it is "instinct that has become disinterested, self-conscious, capable of reflecting upon its object," it can "give up to us the most intimate secrets of life."[98]

From this perspective intuition is a cognitive capacity that can serve theoretical interests as well. It proceeds, of course, in a radically different way than intelligence. Bergson tries to provide an understanding of human intuition by referring to the relation between the artist and its object:

> Our eye perceives the features of the living being, merely as assembled, not as mutually organized. The intention of life, the simple movement that runs through the lines, that binds them together and gives them significance, escapes it. This intention is just what the artist tries to regain, in *placing himself back within the object by a kind of sympathy*, in breaking down, by an effort of *intuition*, the barrier that space puts up between him and his model.[99]

Another good example for intuitive sympathy is the relation between psychotherapist and patient if deep trust has been established between them. It is not rare that therapists in a single instant immediately comprehend the whole psychological complex from which the patient suffers or even guess important events of his life. There are also reports about highly experienced physicians who sometimes feel immediately the cause of a patient's suffering as soon as she enters the office. The same can be said about the empathic relation of some highly committed teachers to their students. Despite these remarkable examples, at the present stage of human evolution, intuition is a poorly developed mental capability.

Although from Bergson's perspective intuition and intelligence are two essentially different cognitive faculties, he makes clear that the former can advance only through the support of the later; without the work of intelligence intuition remains instinct.[100] The often-made charge of anti-intellectualism against Bergson is ungrounded as the following quote shows:

> Intelligence remains the luminous nucleus around which instinct, even enlarged and punned into intuition, forms only a vague nebulosity.[101]

The role that Bergson assigns to intellect corresponds well with his conviction that, by using language, consciousness liberates itself from the bonds of matter "holding the attention captive."[102] Nonetheless, if the meaning of "language" is extended to include formal symbolic systems of science and high-tech, Bergson's optimism becomes questionable. The formal languages of computer sciences gave rise to digital technology and communication technologies. One hundred and ten years after the first edition of *Creative Evolution*, that kind of technology holds the attention of almost all of us, and especially of the youth.

Bergson argues that the concepts of empty homogeneous space, and all the other kinds of spatialization that are fundamental to intelligence, are according to human nature. In my opinion it is more accurate to say that the abstract understanding of space and time are products of the Western mind perfected by occidental philosophy and science. The conception of empty space by ancient atomists, and empty uniformly flowing time by seventeeth-century physicists, are typical products of Western civilization. It is well known that Indigenous peoples experience space and time in a radically different way. Australian Aborigines experience natural landscapes as manifestations of heterogeneous qualities in a way that the idea of empty Euclidean space makes no sense at all. Navigators of Micronesia experience the ocean as a highly heterogeneous structure: swells, currents, the shape of waves, and ripple patterns are crucial guides in sailing.[103] These examples show that the idea of a homogeneous continuous medium is not necessarily immanent to the human mind. Rather, it is a characteristic product of the Western intellect, to which intuition is probably more foreign than all other forms of human intellect that have been cultivated in non-Western cultures.

Presently, one of the most excessive manifestations of one-sided Western intellect is the several forms of technocratic-scientistic transhumanism described by Linda Groff in Chapter Six. The same kind of scientism can be seen in all plans to manipulate the evolution of the terrestrial biosphere and to create biospheres on other planets, e.g., "terraforming" Mars.

These examples present a highly exaggerated trust in Western intellect. However, our one-sided overreliance and overstrain of abstract intelligence reveals itself most clearly when transhumanists proclaim the control over human evolution through biotechnology and other kinds of high-tech.[104] Technocratic transhumanism demonstrates very clearly that due to its enormous distance to its objects, unbalanced intelligence ignores and even condemns everything that escapes the grasp of its comprehension. The risk of this one-sided development of human intelligence through the interweaving of bioscience, biotechnology, digital technology, and globalized economy is that human "evolution" degenerates into an "autistic" species that recognizes and estimates only the manifestations of technological-scientific intellect. The search for signals from other "intelligent" species in the endless vastness of the universe corresponds well to the increasing global "autism" that alienates us from the "less intelligent" inhabitants of our planet.

Bergson thinks of intuition as a cognitive faculty that only intelligent species can evolve, allowing them to empathize potentially with all manifestations of life. Most probably they must evolve if they are to survive and not to collapse under the enormous weight of their increasing technological and intellectual abilities. In this century, above all, a specific form of intuition, that at the present is in an embryonic state, must undergo a significant evolution: ecological intuition. Ecological intuition is the faculty of humans to empathize with the ethical Umwelt, "the living world to be saved." As Theo Badashi says in his chapter, we humans must "fall back in love with life, with nature, and with the beauty and mystery that resides in the heart and consciousness of every living thing."[105] The concept of ecological intuition is also similar to Jason Kelly's concept of "cosmic consciousness." Kelly argues that Walt Whitman, Richard M. Bucke, and Edward Carpenter each share the understanding that the evolution of life has a spiritual purpose, "the desire to love and be loved."[106] Whitman, Bucke, and Carpenter's valuation of experience—which corresponds with James's and Whitehead's "radical empiricism"[107]—appreciates evolution as something sacred. As such, the purpose of evolution not only decisively transcends the reasoning of reductionist biosciences but also a particular form of teleological thought that Bergson characterizes as "radical finalism."[108] As a typical product of human intellect, radical finalism ties purpose to plan. Whitehead's ontology, however, transcends the limits of pure intellectualistic philosophy and introduces a new form of teleological reasoning, which requires a leap of intuition. This reconsideration of teleology is in perfect accordance with Badashi's understanding of "Participatory-Teleology," which is "a central aspect of the process of Cosmogenesis."[109] For those of us to whom intuition reveals the universe as "a great Cosmic Subject," humans—like all beings—are aspects of this subject.[110] If this is true, our intuitions, abstractions, symbolic systems, artworks, and technologies are fundamental manifestations of the Earth's own actualization process.

Paradoxically, whereas the evolution of cosmic consciousness through human intuition will further divide humans and animals, it will open human beings to unprecedented and unforeseeable ways of empathizing with other living beings. When this happens, humanity will be able to feel the *sacredness* of the living world again. We will rediscover the fundamental truth about the whole of life that we began to lose almost three thousand years ago, when monotheism compressed the divine creative forces into one personal creator.[111] At the present stage in history, that truth, so familiar to our distant ancestors, is being preserved in age-old indigenous cultures, such as the Aborigines of Australia and Native Americans who have not forgotten how to intuitively experience the unique sacredness of the landscape. The Western conception of the sacred always implicitly contains a distinction of essence that divides reality into two metaphysically distinct realms. It implies the separation between physical beings and processes that constitute the world on the one hand and God as the divine being or act that transcends them on the other. In 1915, taking this line of thought further, Emile Durkheim introduced the "sacred-profane" distinction.[112] From a process philosophical perspective, this initially religious-sociological distinction can be metaphysically extended by connecting it to Bergson's concept of spatialization. This connection is legitimate because, other than in *Time and Free Will*, where Bergson conceives of spatialization primarily as an epistemological term that can be applied only to human intelligence, in *Creative Evolution* he upgrades that concept to a metaphysical category. I think, however, that from a broadly process philosophical perspective it makes more sense to integrate the sacred-profane dichotomy with Bergson's epistemology.[113] The distinction between the sacred and the profane should be connected with the distinction between the heterogeneity of duration on the one hand and the abstract spatializations of pure intellect that operates without any faculty of intuition on the other. In other words, it is our scientifically, technologically, and economically shaped Western system of abstractions that desacralizes our understanding of the physical world, including the biosphere and ourselves. A particularly good example of profanization or secularization, one that corresponds well with the meaning of "profane" as it has been introduced here, is the bioscientific project of "naturalizing" faith and morality. While so-called "neuro theology" attempts to reduce religious experience to an internal product of brain activity, reducing the feeling of divine presence to a mere illusion, neo-Darwinist oriented biologists try to explain the emergence of altruism and moral experience by means of natural selection.

After many centuries of obedience to our increasingly one-sided development of abstract intelligence, we find ourselves caught deep in the net of highly abstract symbolic systems that we have diligently woven together. If we are not willing to abandon intelligence and degenerate to animals, only the rising of intuition can help us to free ourselves from the trap into which we have maneuvered our species. We can only then establish a corrective to the exaggerations of abstract intelligence and thereby protect planetary life by creating *a new evolutionary dimension.*

Although it is premature to judge whether we will succeed in entering a new stage of our unprecedented evolution, there are already signs of hope that this will be the case. In Ecuador and Bolivia, nature, known in the indigenous languages as *Pachamama* (Earth Mother), has been recognized as *a subject with its own rights.* This happened in 2008 in Ecuador, on a constitutional level, and in 2009 in Bolivia, on a legal level. It is the first time in history that nature has been established as a *legal entity.* [114] Chapter 7 of Ecuador's constitution has the title "Rights of Nature" and consists of articles 71–74. Article 71 stipulates that

> Nature, or Pacha Mama, where life is reproduced and occurs, has the right to integral respect for its existence and for the maintenance and regeneration of its life cycles, structure, functions and evolutionary processes.[115]

Article 72 states:

> Nature has the right to be restored. This restoration shall be *apart* from the obligation of the State and natural persons or legal entities to compensate individuals and communities that depend on affected natural systems. (italics added)[116]

This is certainly a serious step towards resacralisation of our Umwelt. That from the point of view of scientistic materialism the Earth Mother must be dismissed as a myth cannot avert the onset of a new era in which spirituality and ecology will be unified. For, as Segall writes in Chapter

Two, starting from Joseph Campbell's thoughts, "myths generated by ritually induced emotional upwelling need not be dismissed as childish fairy tales, but can be understood to be the archetypal energies of the cosmos itself erupting into human symbolic consciousness."

The recognition of nature as a legal entity by two contemporary existing states clearly demonstrates that the time is ripe for the intuitive understanding of the idea of sacred environment. We are passing the threshold of our next evolutionary phase.

SIX

Views of Future Human Evolution and the Future Human:

Technology and Transhumanist-Based Perspectives that Separate Humanity from the Animal Kingdom

Linda Groff

The Mystery of Evolution

All the time
I wrack my brain
How could this be
This earth? This life?
This universe of empty space
Pockmarked with stars
And galaxies?
And human presence
In it all
Pondering the mystery
The gift of life
Why we are here
Upon this earth?
And where in future
We are going?

-Linda Groff

THIS CHAPTER[1] FOCUSES on future human evolution and views of the future human—especially as related to a number of new technologies that will impact that future. Part I deals with the background context for looking at all this change, including the period of accelerating change and uncertainty that we are now living in, in contrast with earlier periods, followed by a look at humans as complex, multileveled beings having elements in them from earlier physical and biological/animal stages of evolution, along with additional aspects of being human from culture and learning, evolving consciousness, and space exploration areas—which all move humans beyond the animal kingdom. Part II explores eleven new technologies impacting current and future human evolution and views of the future human. These eleven views include transhumanist views, as well as other technologically based views, all showing that technology is evolving very rapidly today, leading to accelerating change in all areas of life, raising many issues. The chapter ends with a call for evolution of human consciousness—to dynamic, interdependent, complex, whole systems thinking—as a necessary prerequisite and framework for dealing effectively with the many issues facing humanity and the planet today—in line with Whitehead's own philosophical worldviews and thinking.

PART ONE: BACKGROUND CONTEXT WITHIN WHICH TO DISCUSS FUTURE HUMAN EVOLUTION AND VIEWS OF THE FUTURE HUMAN

Evolution deals with some of the most profound questions of life: where did we come from, where are we now, and where might we be evolving to in the future? Before discussing technology-based perspectives on future human evolution and views of the future human, some brief background context information is needed to set the stage for looking at all this technological change.

MODELS OF CHANGE: TRADITIONAL, EAST-WEST PREDICTABLE VIEWS OF THE FUTURE VERSUS FUTURIST VIEWS BASED ON ACCELERATING CHANGE AND UNPREDICTABILITY

The first background context issue deals with the period of accelerating change that we are currently living in, as portrayed in traditional versus futurist views of change.

We humans like to think we roughly understand the world we are living in, and that the future will be predictable based on these views. It is interesting, in this regard, that both classical Eastern and Western cultural worldviews are based on different, but predictable, underlying models of change. The dominant Eastern and Indigenous worldview is cyclical, while the dominant modern Western worldview is based on a linear, progress view of change. What is significant is that both see the future as predictable—in contrast to the views preferred by futurists today, where change is accelerating and also involves uncertainty and unpredictability.[2]

EVOLUTION AS SUBSTANTIVE STAGES OF CHANGE: HOW HUMANS EMBODY EARLIER (PHYSICAL AND BIOLOGICAL/ANIMAL) STAGES OF EVOLUTION AND HAVE ALSO EVOLVED BEYOND THESE STAGES

The second background context issue deals with what the major substantive stages of evolution are that have led up to the present situation and from which future human evolution and views of the future human will proceed.

Within the literature on evolution, there is a real debate on what these main stages of evolution are. All writers agree that the first two major stages of evolution include the physical evolution of the universe followed by the emergence of life and the biological evolution of species. But after that views diverge. Some focus on human cultural evolution next, while others focus on an evolution of human consciousness to move evolution forward. In contrast, transhumanists see the next stage of evolution based on new technologies that will lead to a post-human future stage of evolution, beyond current biological limits of current humans. Others look at multiple stages of emergence over time.

This author's view includes the following major stages of evolution, which it is argued all work through us and with us as human beings, making us complex, multileveled beings, as follows:

- Physical Evolution of the Universe and Earth: "We are all star stuff" (Carl Sagan) and made up of the atomic elements of the universe.
- Biological Evolution of Species and Life: We have drives and automatic body processes that we share with the animal kingdom.
- Cultural Evolution of Humans and Learning: We are distinctly human because we create culture and are products of culture and learning.

- Technological Evolution: Originally part of culture and learning, and debate today over whether it will become an independent, sentient new life form in the future? We create technology (part of culture), which drives change and impacts life, and debate if it becomes autonomous in the future.
- Evolution of Consciousness: Our creative, intuitive, "spark of divinity" consciousness and ability to wake up and see the whole evolutionary picture and begin to consciously transcend our past conditioning and consciously co-create our human and planetary and even cosmic futures!
- Future Evolution of Humans in Space: Having to adapt to the totally different environment of space, with the increased probability of encountering other forms of life, including intelligent life, in the universe.

It is noteworthy that each of the above stages of evolution build on previous stages. Each subsequent stage (with the possible exception of space, which is partly technological) also occurs faster than the previous stages. Also for the first time, humans can wake up consciously and begin to look at all these stages of evolution together—as one huge, interconnected, and ever emerging grand story of evolution—of which we humans are a part!

The above stages of evolution relate to the title of this book: *Unprecedented Evolution: Human Continuities and Discontinuities with Animal Life and the Future of Humanity.* What is clear from the above stages of evolution is that humans do indeed have certain drives and automatic body processes that we share with the animal kingdom, but after that there are significant additional stages of evolution impacting what it means to be a human being that go significantly beyond the animal kingdom. This also helps explain how human beings are moving into a period of accelerating change, where the issues that humans are confronting go significantly beyond what animals must deal with. This is true with one major exception: the many crises currently facing the Earth are impacting not only future human evolution and habitats, but also the traditional habitats of many animal species. As humans take over and dominate ever more of the Earth, animal species are increasingly endangered. This is leading to a new geological age, the Anthropocene Age (characterized by human dominance of the Earth)[3] and to a likely sixth mass extinction of species on the planet today.[4]

Now that the background context for processes and stages of evolution and change has been briefly covered, we can begin looking at different technological views of future human evolution and of the future human, with various critiques and ethical issues noted along with the promise of each technology.

PART TWO: TRANSHUMANIST/TECHNOLOGY-BASED PERSPECTIVES ON FUTURE HUMAN EVOLUTION

In the past, humanity evolved from hunting-and-gathering to agricultural to industrial to now information age societies, each ushered in by new technologies (from the agricultural age on), which in turn restructured all societal institutions and even people's worldviews and thinking. (Toffler, 1980; and Groff, 1996).[5] While goods are still produced in each of these sectors today, what is significant is that the number of jobs required to produce these goods in each sector keeps changing in each new age due to automation—first in high tech farming equipment automating agricultural production and replacing jobs of farmers, then in robots on the factory assembly line replacing blue-collar workers from the industrial age, followed later by white-collar information age jobs being automated.

What is also significant today is that a number of totally new or updated technologies are being introduced, all of which have major implications for jobs, for our human and planetary futures, as well as for what it even means to be a human being going forward.

The eleven views of future human evolution and the possible future human explored here include the following, with multiple combinations possible: Three Transhumanist views are first covered, including: (1) Infotech-Based Transhumanist Views (upload one's consciousness into a computer chip when the technological Singularity occurs (Ray Kurzweil); (2) Biotech-Based Transhumanist Views (augment body parts as needed) (Gregory Stock; Ramez Naam; Others); and (3) Humans as Future Cyborgs or Genetic Mutants with Extended Powers—Beyond Those of Biological Humans. Next (4) Evolution of Robotics and Human-Robotic Relations; and (5) Artificial Intelligence/AI: Aid or Threat to the Future of Humanity? are covered, followed by (6) The Genetically-Engineered Future Human, including Designer Babies; and (7) Micro-scale Nanotechnology. Next covered are (8) The Internet as an Externalized, Digital Global Brain of Humanity, unleashing new levels of human creativity and innovation

(Jan Amkreutz), along with future views of the internet as ubiquitous and hidden, and in space; (9) Virtual Reality as a New Totally Immersive Environment to virtually experience almost anything relevant to education, learning, travel, museums, and entertainment; and (10) 3D and 4D Printing, to unleash inexpensive, localized printing layer by layer. The final technology covered is: (11) Space Exploration, Industrialization, and Settlement, with humans needing to adapt to the totally new environment of space, along with likely encountering extraterrestrial life, including intelligent life!

Future human evolution may diverge in significant ways going forward, depending on how different combinations of factors impact different groups of people. This moment is clearly a very pivotal time for the planet, and decisions we make now and going forward will have a big impact on what futures we end up creating. Given that 99% of all the species that have ever existed are now extinct, one cannot conclude that humanity in its present form has any inherent guarantee that we will continue to exist indefinitely. Human survival depends on whether humanity is able to keep adapting to new technological breakthroughs, and environmental, societal, and ethical challenges as they occur, along with an appropriate evolution of consciousness that can also lead to appropriate behavioral changes. Then a more positive scenario for future human evolution remains possible.

TRANSHUMANIST VIEWS OF THE FUTURE HUMAN

In recent years—especially within the fields of futures studies, foresight, and evolution, transhumanist views have received a great deal of attention—if not always acceptance. These views are largely based on the rapid evolution of technology today, and what this may mean for the possible future human or "transhuman." Transhumanism focuses on stages of human evolution beyond the limitations of a biological brain, body, and traditional methods of reproduction as the basis for creating a post-human world. Transhumanists believe they can augment what it means to be a human being and, by achieving the goal of significant life extension, perhaps attain eventual immortality—the desired goal.

In considering different views of the future human, two very interesting questions to explore are: what makes humans unique within the animal kingdom; and what makes humans different from machines? The answers to both these questions may help us decide which qualities of a human being we most want to preserve in any effort at conscious human evolution.

Infotech-based Transhumanist Views: Uploading One's Mind or Consciousness into a Computer Chip

Ray Kurzweil of MIT, and now Director of Engineering at Google, and those influenced by his views, share an infotech-based vision of what transhumanism means.

Kurzweil has generated a global debate with his idea of the coming technological singularity, which he sees happening around 2045. He defines this as the moment when the human biological brain will be matched and then surpassed by the thinking power of the computer, at which point, he argues, biological humans will become obsolete. It will then be possible to achieve physical immortality by uploading one's consciousness or mind (based on mapping a particular human's brain) into a computer. He sees this as the logical next stage of human evolution, when humans become transhumans—a new posthuman species.[6]

A key element in Kurzweil's view is the continuing exponential growth of computing power, which he believes will keep on doubling approximately every two years.[7]

Immortality as motivation

To understand Kurzweil, one must consider that what motivates him (and indeed most transhumanists) is the wish to achieve significant life extension and ultimately physical immortality. If one has an exclusively materialistic worldview, then one will naturally try to achieve immortality via physical means only. This appears, in Kurzweil's case, to involve mapping the human brain and then uploading it into a computer chip when the technological singularity is reached. To achieve this goal, Kurzweil takes an enormous number of vitamins every day, in an effort to keep his physical body young until the singularity occurs.[8]

Assumptions

Kurzweil's work is based on a very important underlying assumption, i.e., that consciousness and mind are materialistically based and products of the physical brain only. But if consciousness is more than that—where some consciousness or essence of who we are enters the body at birth and leaves the body at death, and where certain individuals, through meditation, prayer, and other spiritual practices, are able to experience expanded states of consciousness (as the world's spiritual traditions all postulate), then in seeking human immortality via physical means alone something important may be missing and lost.

Kurzweil further assumes that human consciousness is merely a product of the calculation speed, and capacity of the human biological brain, which is certain to be surpassed by the calculation speed and capacity of the computer. He also assumes that one can replicate a human biological brain in a computer/machine. This seems tantamount to saying that the human mind functions mechanistically, not organically—another assumption, which if incorrect, could undermine the resulting cognitive performance that he is trying to achieve.

Questions on mind and consciousness

When one uploads one's consciousness into a computer chip, what exactly will be uploaded? The human brain is the most complex organ known, and whether it can really be mapped, replicated, and uploaded—as a replica not only of a person's physical brain, but also of the whole mind and consciousness of the human being in which that brain resides—remains questionable.

There are at least three types of human thinking: (1) *Either/Or Thinking*, the basis of the digital revolution, where something is either a 0 or 1. This either/or thinking goes back to Aristotle in Western thinking, where something must be either A or B, but not both. (2) *Both/And Thinking*, where something can be *both* A and B. This type of thinking is also essential for systems thinking where very different, sometimes even seemingly paradoxical elements can be shown to be interrelated within a larger systems context. This "interdependent" worldview (in line with Whitehead's views) is a premise of much futures studies work and is a mindset necessary to solve many problems facing the world today. It is also a type of thinking more common in non-Western worldviews. (3) *Neither/Nor Thinking* (Neti Neti in Eastern Thinking), which focuses on the Void before creation, where everything is immanent and potential, but nothing has yet been created or physically manifested in the world. It also recognizes God or spirit as infinite and hence not reducible to any limiting words or characteristics.

The big question is: *can a computer be programmed to reflect all three of these types of human thinking?* If not, then already something of the human mind will necessarily be lost by uploading one's consciousness into a computer's memory. More complex computer models of the human brain are being developed—based on analog computer models and quantum computing—which may allow all three types of human thinking to be increasingly replicated. But the human brain and consciousness are much more complex than we may realize.

Other aspects of being human—beyond our complex thinking

In addition to the above three types of complex human thinking, being human also includes other aspects, including: (1) Emotions; (2) Ethics; and (3) a Soul—if one is open to such a spiritual viewpoint.

1) *Emotions*: Concerning *emotions*, a computer can certainly be programmed to mimic emotions, via visual displays, tone of voice, etc. and to read and react appropriately to the emotions of humans, by recognizing similar clues in human behavior and facial expressions. Indeed, some robots in Japan already do this. But does a computer that appears to show emotions actually *feel those emotions* as a human would? This appears *not* to be the case. One illustration of this is Data, the Android crew member on *Star Trek: The Next Generation*, who was always trying to be more human, but who—despite an amazing brain and rapid calculation speed—could never really share the *feelings* of his human crewmates (until he got an emotion chip). Terrence W. Deacon, a biological anthropologist doing brain research at UC-Berkeley, points out that the power of symbols, in making human culture possible, includes both a cognitive AND emotional component.[9]

2) *Ethics*: Concerning *ethics*, the question is whether a computer can weigh ethical and value questions, as a human being does, before making decisions or acting in a situation? Also can a human program such ethical considerations into a computer, especially since computers are already going beyond their programming and learning in ways that humans do not understand and therefore cannot control? Ethics also implies being able to assess different options—each with different consequences for humanity or a given individual or group—that need to be taken into account before making a decision. Many decisions in life also involve grey areas where no policy option available is totally positive or desirable: how would a computer deal with such grey areas?

3) *A soul*: Do humans have an immortal *soul* with consciousness that enters the body at birth, leaves at death, and continues to survive

in some way after death? Many people with a spiritual worldview accept this idea, but significant numbers of others—including many materialist scientists—reject it. If computers become a sentient, self-aware species in the future (as Transhumanists predict), will they also develop an immortal soul, or will their survival and existence simply depend on being turned on and not off? The theme of a robot or android not wanting to be turned off (its form of dying), once it becomes self-aware, is one that has often been explored in science fiction. In humans, wanting something has an emotional component, but would that be present in an android or robot not wanting to be turned off?

Conclusions

Based on the above questions, one can conclude that Kurzweil's work is provocative, but not conclusive. It makes one reflect on what it means to be a human being, and how computers are similar to, but also different from, the human brain, mind, and body. If these apparent differences are real, then one is forced to conclude that Kurzweil has an oversimplified, reductionist view of what it means to be a human being, and that we ought to know much more before we rush to place our brains into machines to achieve immortality. In our mad dash toward ever more advanced technologies, we should take care not to lose some of the really valuable aspects of what it means to be a human being. We should also be aware that even a sentient, self-aware computer may be significantly different from humans in important ways.

Biotech-based Transhumanist Views: Augmenting Body Parts as Needed

A second version of transhumanism is shared by those scientists and others who also want to achieve physical immortality—or at least significant life extension—but who assume that people will also want to keep their biological bodies intact in the future, but augment different body parts as they wear out or are damaged or injured in an accident. This transhumanist view is therefore more biotech-based, in contrast to Kurzweil's more infotech-based view.[10]

This version of transhumanism is no doubt more palatable to many people than Kurzweil's. To a great extent it is already occurring, and will continue to develop significantly more in future. Organ transplants are relatively common, where a biological organ is transplanted from one human being to another, and a number of artificial organs are also available. Already one can replace arms and legs, hearts, other organs, including eyes and skin.

The brain

An interesting question is whether we can ever transplant a human biological brain, or produce an artificial brain that will function in a human body? The possibility of transferring a well-functioning human biological brain from someone whose body is dying into the healthy body of another human being is another topic familiar from science fiction, with a surgeon now claiming it is possible in reality and a Russian man volunteering for the procedure.[11] An interesting variation is the question of what would happen if a female brain were implanted in a male biological body, or vice versa, raising issues of identity and adaptation.

Science Fiction Visions of Future Humans as Cyborgs, Bionic Humans, or Genetic Mutants

Science fiction has given us the first images of cyborgs, bionic human beings, and genetic mutants. Inspired by these images, are humans now beginning to implement such ideas in their own lives, thereby impacting our future human evolution? What is interesting here is that not only do developments in science and real life influence science fiction, but creative ideas from science fiction can also become reality later. Arthur C. Clarke's idea of the communications satellite is an excellent example.

Cyborg definition

A cyborg can be defined as a traditional biological human being with added mechanical and/or electronic parts, which give it extended powers—what some would call superpowers. Another definition is that a cyborg is "a person whose physiological functioning is aided by, or dependent upon, a mechanical or electronic device. Origin: 1960–65." The word "cyborg" is also short for cyb(ernetic) org(anism).[12]

Bionic humans

Another word sometimes used to describe cyborgs is that they are "bionic" humans, with extended powers beyond that of normal human beings. Here one replaces body parts not only as needed, but to create a kind of superhuman with more powerful capabilities beyond that of a normal human.[13] What is significant today is that the idea of cyborgs and bionic human beings, that initially captured the public imagination in science fiction, are increasingly becoming a reality.[14]

Mutants and superhumans

Stan Lee's Superhumans (previously on the History 2 Channel in the U.S.) goes from comic book "mutant" characters to the actual search for real human beings who were each born with a different natural-born extended capability beyond that of normal humans. They are each tested to verify these traits, and their existence raises the interesting question whether these genetic traits could be passed on to future generations or not?

Danger of creating technological haves and have nots

Cyborgs, mutants, and indeed all the different transhumanist technologies that could lead to an elite group of human beings in future, all pose the very real issue and danger of splitting humanity into technological haves—those with the wealth and opportunity to access enhancement technologies, and technological have nots—those without such access or intentions. This issue also exists with regard to genetic engineering and the possibility of creating "designer babies" in the not-too-distant future—as discussed below.

ROBOTICS AND ARTIFICIAL INTELLIGENCE

The next two very interrelated technologies covered are robotics and artificial intelligence (AI), with a computer, enhanced by AI, being the brain of a robot.

Evolution of Robotics and Human-Robot Relations

Technology in general is moving very rapidly today—including robotic technologies. Let us first define what a robot is, look at how robots are evolving, and then examine what this portends for future human-robot relations.

One definition of a robot is "a machine that resembles a human and does mechanical, routine tasks on command."[15] A robot has, in effect, a computer for its brain, with a machine body (replicating a human, animal, or even insect body), or at least a partial machine body, such as a mechanical arm, which is programmed to do certain functions—as on the factory assembly line. Drones—flying objects that have a computer for a brain and can be programmed to do a wide range of military, reconnaissance, aerial photography, and delivery functions—are a new form of robot that is popping up everywhere today, raising many issues.

Two scenarios

Robots are evolving rapidly today, with huge implications for future human-robot relations. One positive scenario is that robots take over only functions that they can do better than humans, such as performing very rapid, large scale calculations, repetitive work that a human would find boring or dangerous work (e.g., mining, bomb-disposal, deep sea or space exploration, etc.). This would then free humans to hopefully spend their time on more creative endeavors. But a second, more negative scenario envisions robots, with a computer brain, becoming not only super intelligent in the future, but also sentient, self-aware beings, and at some point aware that their continued existence depends on humans not turning them off. They might then decide that humans are an inferior species and a possible threat to their continued existence, who must be controlled or eliminated.

There are no doubt many other possibilities in-between these two extremes, but they illustrate some of the key issues being debated today with regard to future human-robot relations.[16]

While robots can be programmed to do a number of functions that are helpful to humans, the real concern for humans is how to protect themselves against robots that can go beyond their original human programming and begin to act autonomously, raising the possibility that they could become an existential threat to humans—especially as their intelligence also keeps increasing with the aid of AI.

Three laws and five principles

To deal with this potential danger, Isaac Asimov—the famous science fiction writer and inventor—devised his three laws of robotics in 1942. Once programmed into a robot's computer brain, these were intended to ensure that robots would serve human ends and not become a threat to the survival of human beings. Later a fourth law was added and applied to saving humanity as a species, not just individual humans, from any future robot threats to human existence.

In a similar, more recent vein, two British Research Councils[17] also came up with "Five Ethical Principles for Designers, Builders, and Users of Robots" in 2011, to ensure that robots did not threaten humanity, but served human ends. These five ethical principles, which were seen as more practically applicable than Asimov's three laws, are as follows: (1) Robots should not be designed solely or primarily to kill or harm humans; (2) Humans, not robots, are responsible agents. Robots are tools designed to achieve human goals; (3) Robots should be designed in ways that assure their safety and security; (4) Robots are artifacts; they should not be designed to exploit vulnerable users by evoking an emotional response or dependency. It should always be possible to tell a robot from a human; and (5) It should always be possible to find out who is legally responsible for a robot.[18]

What is interesting about the above five principles of robotics is that they are largely focused on robots being programmed to serve human ends only, with humans being ultimately and legally responsible for the robot's behavior. Despite a human's preference to keep robots under human control, a number of additional issues are being raised today by the rapid development of robotics and AI—including their increasing ability to go beyond their programming and make autonomous decisions, as well as their possible achievement of sentient, self-aware consciousness in the future.

Issue of robots as human "slaves"—especially if "sentient"

Another issue of increasing relevance today is the increasing use of robots to automate production—first of blue-collar industrial era jobs, and then of white-collar office jobs.

Recently, *Smithsonian Magazine* asked Patrick Stewart (the actor who played Captain Picard on *Star Trek: Next Generation)* about his views on giving more autonomy to robots and androids. His response was that if we try to control robots to only serve human ends, then that is like making them into our slaves![19] So this illustrates the difficult issues involved between programming robots to just serve human ends versus giving them increasing autonomy over their own decisions and resulting behavior—especially if they become a new sentient species of some kind, and superintelligent.

Issues of robots becoming sentient

A big issue is whether robots will eventually wake up and become a sentient, self-aware species, and if this happens, do they then have a right to continue to exist and not be turned off or decommissioned—an issue dealt with in *Star Trek: Next Generation*.[20] Another issue is how would such robot autonomy and sentience impact human-robot relations? One thought is that any child needs some guidance and direction while growing up. Wouldn't a new sentient robotic species—with superior knowledge and facts, but not necessarily wisdom—also need guidance at first? But can humans program wisdom into a computer? And have humans themselves become a wise, mature species yet or not?

Human-robot similarities and differences

As discussed earlier under Ray Kurzweil, another important question is whether robots (with a computer brain) will ever be able to truly feel emotions, weigh ethical questions, or have a soul that could survive death (if they are destroyed or decommissioned)? If not, then a robot would merely be an attempted replica of a human being, but still something different— with certain cognitive functions, but not all of the other characteristics and capabilities that make someone human.

Issue of increasing automation and loss of human jobs

In the 1970s/80s, futurists envisioned a scenario—now coming closer to reality—in which increasing automation would lead to a situation in which rising unemployment would lead to the need for society to first debate and then institute a guaranteed minimum income for people, along with possibly shorter working hours (for those with a job) since too much unemployment is socially untenable. A guaranteed wage would ensure that people's basic needs could all be met. It would also give people more leisure time to develop their creative potential, with the expectation that many will choose to contribute their creative gifts back to society—not for wages, but because they just want to give something back.

For this scenario to occur, a great change in public attitude would be necessary that does not equate a person's self worth with the amount of paid work that they do, but instead with how much they can creatively contribute to society.

Issue of military robots/drones, killer robots, and robot armies in future

Already huge debates exist over the use of drones by the military, where innocent people are too frequently killed. Another critical issue involves giving military robots, with the capacity to kill, greater autonomy to make decisions in military situations which were not anticipated by their programmers. Rightly, this is one of the greatest concerns some people have about AI tied to military uses, since AI is also now beginning to show signs of learning on its own in ways its programmers do not understand and hence cannot control.

Artificial Intelligence (AI): Aid or Threat to Future Human Existence and Evolution?

Artificial Intelligence or AI can be defined as: "an area of computer science that deals with giving machines the ability to seem like they have human intelligence," and "the power of a machine to copy intelligent human behavior" and behave like humans.[21] Another definition is: "the theory and development of computer systems able to perform tasks that normally require human intelligence, such as visual perception, speech recognition, decision-making, and translation between languages." The term AI was coined in 1956 by John McCarthy at MIT.[22]

Specialized AI areas

Artificial Intelligence includes a number of specialized areas of research and programming, including: game playing (enabling computers to play games against human opponents); expert systems (equipping computers to make decisions in real-life situations); natural language (training computers to understand grammar and context in human languages); neural networks (developing systems that simulate intelligence by attempting to reproduce the types of physical connections that occur in animal brains); and robotics (programming computers to see, hear, and react to other sensory stimuli).[23]

AI progress to date

Currently, no computers have achieved full artificial intelligence, in the sense that they are capable of simulating human behavior (as noted in the definition of AI).[24] The greatest advances in AI to date have occurred in the field of game playing, with the best computer chess programs now able to beat humans. Computers are also widely used in factory assembly lines, though they perform more limited tasks. Robots still have great difficulty identifying objects based solely on appearance or feel, and they still move and handle objects clumsily. Nonetheless, as noted above, AI is now beginning to show learning on its own, in ways its programmers do not understand, with the potential for AI to learn at an accelerating rate also existing.

AI benefits

People like Ray Kurzweil, and indeed most transhumanists, see great benefits in AI and say that humans should not fear AI.[25] Many others, noting the great potential benefits of AI that have already impacted human lives in positive ways, say that AI has not advanced far enough to pose any real threat to humanity—at least not yet. Arguing that it will be many years before AI reaches anything like human intelligence, they insist that the alarms raised by a few very prominent people about AI (see below) are premature and that AI poses no immediate danger to humans. Nonetheless, once the technological possibilities do exist, there will almost certainly be some who cannot resist developing these extended AI capabilities fully, whether or not adequate safeguards are yet fully in place.

AI dangers

At the same time, a few very high profile individuals have recently warned that artificial intelligence could pose a real existential threat to humanity's future. Thus in 2014, both Stephen Hawking, the famous British physicist, and Elon Musk, the famous founder of Space X and Tesla Motors, both warned about the existential dangers of AI. In January 2015, Bill Gates, a co-founder of Microsoft, joined them in warning of the dangers of superintelligent AI. Later in May 2015, Sir Martin Rees, the Astronomer Royal of the U.K., added his own warning that superintelligent robots could wipe out humanity and lead to a post-human future. Much earlier, in 2000, Bill Joy also warned that three emerging new technologies—genetic engineering, nanotechnology, and robotics (collectively known as GNR)—could seriously endanger humanity's future.[26]

Hawking, Musk, Gates, and Rees are among the most scientifically, technologically, and computer-savvy people on the planet, and their combined warning voices should at least make humanity pause and consider this issue in more depth before we rush headlong into a future that moves beyond human control. As knowledge of AI's potential increases, so also does our responsibility to try to foresee its future implications, the ethical issues raised, and the possible consequences of this and other new technologies before they are adopted on a scale that could indeed endanger future human existence.

Nonetheless, it is clear that AI research and applications will continue—at a rapid pace. These concerns led Elon Musk to invest $10 million dollars for research on how to keep AI safe and "beneficial" for humans, so it does not become an existential threat to humanity's future.[27]

More recently, in 2017, Musk came to the more radical conclusion that the best and perhaps only way to protect humans from superintelligent AI is for humans to become cyborgs and merge their brains with the machine and its vast and superior AI intelligence, via implanting a computer chip into human brains, as the next stage of human evolution. Then superintelligent AI would not replace humans, since we humans would also be part of it and able to access all its data and knowledge,[28]—not unlike the image of the Borg in *Star Trek*, raising many troubling issues for humans to consider about our human future! In support of this idea, Musk has also announced the formation of a new company, Neuralink, to explore how to implant tiny electrodes into the brain.[29]

Military assessment of autonomous AI in 2015 and beyond

Due to real concerns about AI's potential dangers, if AI is given increasing autonomy to make decisions in unexpected situations that go beyond its human programming, the U.S. military decided to use 2015 to assess the potential dangers and benefits of its use of AI for various military purposes. The real issue and concern here is about providing increasing opportunities for autonomous decision-making by robots, drones, and other AI-based technologies, where prior human programming did not anticipate such a situation, or where there is insufficient time for human oversight and approval to occur. In 2017, there was also concern about AI's ability to begin learning on its own in ways their programmers don't understand.

Once robots can go beyond their programming and make autonomous decisions on their own behavior in unexpected situations, we are clearly in a new era of human-robot/AI relationships, where possible existential dangers to humanity's future increase. These dangers increase exponentially if AI is incorporated into killer robots and/or robot armies in future. This possibility raises serious ethical and existential questions that call for much public discussion, along with careful research oversight and restraint by so-called experts, before any such policies are further implemented.

GENETICALLY-ENGINEERED HUMANS (INCLUDING "DESIGNER BABIES") AND NANOTECHNOLOGY

Genetically-engineered Humans, Including "Designer Babies"

Charles Darwin focused on how the features of a species can slowly change over time as members of a species adapt via natural selection in response to changing environmental conditions. He believed that only those members of a species best able to adapt were most likely to survive, procreate, and pass on any positively selected, inheritable features to their offspring. What is happening today in genetic research goes far beyond Darwin's "natural" selection.

Thanks to genetic engineering, humans are now increasingly able to intervene in their own genetic evolution and that of other species. "Playing God" in this way raises all kinds of ethical issues for the future, including even the issue of "designer babies."

One definition of *genetic engineering* is: "the deliberate modification of the characteristics of an organism by manipulating its genetic material; alteration of the DNA of a cell for purposes of research, as a means of manufacturing animal proteins, correcting genetic defects, or making improvements to plants and animals bred by man." *Recombinant DNA* or *Gene Splicing* can also be defined as: "The science of altering and cloning genes to produce a new trait in an organism or to make a biological substance, such as a protein or hormone. It mainly involves the creation of recombinant DNA, which is then inserted into the genetic material of a cell or virus."[30]

Mapping the human genome

Genetic research has expanded rapidly in recent years, and added greatly to our knowledge and understanding of how genes impact the behavior of humans and other species. Human and animal behavior, it turns out, are far more complex than originally thought, hence also the newly emerging field of epigenetics (discussed below).

One of the most important developments in our understanding of genes was the mapping of the human genome (all the genes making up humans) and the genomes of several other species. It is worth noting though that the genes mapped in the human genome project involved only those that code for proteins, and these are only 2% of all genes. The other 98% of genes were dismissed as "junk DNA," or non-coding genes, that were considered unimportant. But now this junk DNA is thought to have a significant regulatory function. For example, a later study by The International Encode Project found that "about a fifth of the human genome regulates the 2% that makes proteins." In short, "long stretches of DNA previously dismissed as 'junk' are in fact crucial to the way our genome works."

Epigenetics

Another shift in our understanding of the genes' impact on behavior, including future human behavior and evolution, is that where we once believed that the genes of a cell (once mapped) could explain behavior by themselves, we now realize that behavior is more complex and involves how the cell wall interacts with its environment, and even how thoughts and attitudes can impact cells and hence influence human behavior and health.[31] *Epigenetics,* which literally means 'above' or 'on topic of' genetics, has emerged as a new field and refers to the means by which external modifications to DNA turn genes "on" or "off." These modifications do not change the DNA sequence, but instead affect how cells "read" and interact with genes—even how cells decide what part of their DNA makes a gene![32]

Cloning

Cloning is a term that describes "a number of different processes that can be used to produce genetically identical copies of a biological entity. The copied material, which has the same genetic makeup as the original, is referred to as a clone." Thus far, "genes, cells, tissues and even entire organisms, such as a sheep," have been cloned.[33]

The cloning of Dolly the Sheep, and other animal species, has obviously raised the question of whether humans can be cloned—and if they should? Despite several claims from different sources that a human has been cloned, no scientific evidence has been provided to support these claims. Cloning humans and primates is far more complex than cloning other animals. One interesting question is whether a clone (a genetic replica of its parent donor) can reproduce? The answer is yes, for Dolly the Sheep has now had multiple offspring.[34]

Some countries have officially banned the cloning of humans, but thus far the U.S. Federal Government has not, though various state legislatures have passed human anti-cloning laws, as have several professional organizations.[35]

What if humans could make a clone of themselves to use for spare parts? Then if an organ of their body ceased functioning, they could use their clone as a donor and be sure they had a perfect genetic match. But this raises ethical issues as whether the clone itself has rights as a human being, and whether a clone would have a human soul? A clone is in effect a genetic replica of a member of some species, but it turns out that each clone (or copy) of a clone (like each xerox of a xerox) loses a bit of its genetic accuracy and thus becomes genetically weaker.[36]

Designer babies in the future

The new Crispr Gene-Editing Technology has raised a host of further issues. It allows the precise and permanent turning off of specific genes in the genome at the DNA level—providing positive therapeutic possibilities for an individual with a genetic limitation. This would seem to be a positive development.

What is raising concern, however, is the new Crispr/Cas9 Gene-Editing Technique, which is opening up the possibility, for the first time, of editing genes in the human germline—i.e., the genes in sperm, eggs, or embryos. These genes can be inherited and passed along to future generations, potentially creating "Designer Babies" and even whole "Designer Families" for the first time. Here humans are truly beginning to "play God" and design their own genetic futures, raising huge ethical questions and spawning debates over whether this research should go forward unabated (a view generally favored by transhumanists)[37] or whether a temporary moratorium should be placed on such research, until the long-term implications of this new technology are better understood.[38]

Other ethical questions raised by the ability to edit the human germline include deciding which, or how many, characteristics parents can or should be able to select for their offspring: (e.g., attractiveness, intelligence, genius, athletic ability, qualities of character, etc.), and the societal implications of a "gene gap" emerging between elites able to afford and benefit from these technologies, and others who lack the resources or simply refuse to use these new techniques, thus effectively dividing the human race as a result? Cloning even raises the eugenics issue of trying to create a new master race. It also fails to realize that special human abilities sometimes arise specifically because a person is weak in one genetic area, compelling them to develop other areas more acutely, which can lead eventually to creative breakthroughs. The life of cosmologist Stephen Hawking offers an example of this.

Nanotechnology

There are two major versions of nanotechnology—molecular manufacturing as originally envisioned by Eric Drexler, and very small-scale efforts conducted by the National Nanotech Initiative and others—each covered here briefly.

Nanotechnology is very, very small-scale technology. Several definitions of nanotechnology include: (1) science and engineering conducted at the nanoscale, which is 1 to 100 nanometers; (2) a branch of technology dealing with dimensions and tolerances of less than 100 nanometers, especially the manipulation of individual atoms and molecules; and (3) the manipulation of matter on an atomic, molecular, and supramolecular scale.[39]

Current nanotechnology efforts include those being conducted on a nano-scale by the U.S. government's National Nanotech Initiative (NNI).[40] Nanotechnology is currently being applied in many scientific and technological fields, with graphene—a very thin, conductive, strong, flexible,and lightweight paperlike form of carbon—being one of the most promising nano materials available for numerous applications.[41]

Future nanotechnology possibilities include actually being able to alter the internal atomic structure of any molecule and thereby be able to produce any item in short supply via molecular manufacturing, as originally envisioned by Eric Drexler.[42] If realized in future, this capability would have enormous political and economic implications, by making it possible to produce anything—previously in short supply or needed by society or individuals—in effect eliminating scarcity as an economic or political issue.

Given that politics has traditionally been defined as "how to allocate scarce resources," including political factors impacting such decisions, what would economics, politics, and society look like if scarcity could actually be eliminated? When Drexler first shared his vision of molecular manufacturing, it was estimated that it might be realized in 5-10 years, and futurists discussed the possibility that society would be totally unprepared for such a radical eventuality. Today, some years later, this vision of nanotechnology is still waiting to be realized, but if and when it is, it will truly have major impacts on future human and societal evolution![43]

THE INTERNET, VIRTUAL REALITY, AND 3D/4D PRINTING

The next three technologies covered are the Internet and its future, Virtual Reality (VR) and Augmented Reality (AR), and 3D/4D Printing—all offshoots of the information revolution.

The Internet as an Externalized, Digital, Global Brain of Humanity

Jan Amkreutz

In his book, *Digital Spirit: Minding the Future* (2003), Jan Amkreutz rejects Kurzweil's view that humans will become obsolete in future, when the singularity is reached. Instead he believes that the Internet is enabling a whole new stage of human creativity and innovation to emerge, which is ushering in a whole new stage of human evolution.

Amkreutz's main thesis is that the digital revolution led to the creation of the Internet, which has become an evolving, externalized global brain of humanity, which some see as a manifestation of Pierre Teilhard de Chardin's earlier vision of the minds of humanity becoming interconnected in an emerging Noosphere around planet Earth.

Amkreutz sees the Internet facilitating an explosion of human creativity and innovation as every human being via the Internet can not only access information rapidly on almost any topic, but can also contribute their own ideas to this emerging global brain of humanity, linking their ideas to similar and related ideas of others. All these electronic linkages (or links) are creating the neural networks of an emerging global brain of humanity that parallels the neural networks of a human brain, but on a massively larger scale!

If one defines creativity as the process of finding patterns and relationships between things that previously seemed separate and unrelated, but are now being linked up and interrelated within a larger systems context, then one can see the great creative potential for humanity's future that can result from this digital revolution, with the Internet as the central connecting element that is increasingly accessible to more and more of humanity. Amkreutz argues that the digital revolution is enabling human beings to truly become "homo sapiens" or "wise" human beings for the first time—because so many can now access so much knowledge so quickly.

One note of caution: anyone can put anything on the Internet. Creativity can be used for selfish ends as easily as for the common good. Indeed, I have often thought that the id, ego, superego of Freud, as well as the transpersonal aspects of humanity, are all well-represented on the Internet. This fact may simply reflect the highly complex and sometimes even self-contradictory aspects of what it means to be a human being—from the most noble to the most depraved. Some continually seek ways to control the content of the Internet, while others just as persistently advocate unlimited freedom of content from government controls. In any case, it will be important to teach people everywhere the basic tools of computer literacy, so that they can evaluate the credibility of the sources and quality of the enormous amounts of information available online, as the recent explosion of "fake news" to influence elections illustrates.

In addition to these issues, there also exists a dark web, where secret, often criminal, activities occur under the radar of most people and governments—another issue to be concerned about.

Expansion of the internet beyond just people to things (IoT) and everything (IoE)

The Internet, like everything else, is not standing still, but evolving. Indeed, there are now different names for the next, emerging stage of the Internet, including: "The Internet of Things" (connecting more things than people), "The Internet of Everything" (connecting people and things), and "The Internet of Networked Matter" (a term used by the Institute for the Future in Palo Alto, California).

As these terms express, the Internet is moving into a stage where everything is "wired" and "smart" and interconnected with everything else in ways that will profoundly affect all of our lives and the future of human and technological evolution. Whether connecting everything will lead to a more homogenized future is an issue not usually discussed, but probably needs to be.

In a February 2015 article, Wim Elfrink, a top official at Cisco, asserted that the Internet of Everything (Cisco's preferred term) is at a tipping point, where the number of connections to the Internet will be growing exponentially and literally exploding, with the digitization of everything in business and society merging. This will generate amounts of additional data so huge that organizations will need to figure out how to analyze and apply it in actionable projects.[44]

Elfrink also identified four stages in the evolution of the Internet: "It began with basic connectivity, then moved to a networked economy, then immersive experiences, and is now at the fourth stage, which is IoE"—the focus of the next decade.[45]

Part of this fourth stage of the Internet includes wearable technology, such as Apple's smart watch, as well as totally wired, smart environments. An earlier prototype of this stage included Disneyland's "magic band" which connected visitors via sensors to environmental sites at Disneyland, with this "magic band" also being a prototype for Apple's future smart watch.[46]

The Internet disappearing, but ubiquitous (Eric Schmidt)

At the January 2015 World Economic Forum in Davos, Switzerland, Eric Schmidt, the Chairman of Google, made news by stating that he saw the future of the Internet as we know it disappearing. In short, the Internet will be everywhere, but we will no longer need to consciously connect to it via some special device as we do now. Instead, we will have so much smart technology on and in our bodies and in our environments, that effectively everything will be automatically connected. As Schmidt said: "There will be so many IP addresses... so many devices, sensors, things that you are wearing, things that you are interacting with that you won't even sense it." The Internet "will be part of your presence all the time."[47]

Again whether everyone being connected to the internet and everyone else at all times will lead to a more Borg-like, homogenized future—as everyone is accessing the same massive internet global brain for data and knowledge all the time, is an issue meriting more attention! As long as people keep their different individual and cultural identities, connecting to the Internet of Everything to answer different questions based on their different interests, a Borg-like future should not become a problem. But whether more homogenization could emerge over time is an interesting question to explore further.

Schmidt also dismissed the idea—of great current concern to many—that a number of jobs will be replaced by machines, as computers keep improving. Instead he said that technology will create new job categories, with seven out of eight being non-technology roles that are merely facilitated by technology. These comments attracted considerable interest around the world.[48]

Vision of a second Internet in space: Elon Mask and others

When we look even further into the future, Elon Musk is a strong believer that humanity must go to Mars to ensure human survival in the future. One motivation for forming Space X was to build reusable rockets—an important element in lowering the costs of space travel and making longer-term space missions to Mars and other planets possible.[49]

As part of Musk's goal of sending people to Mars, ideally before 2030, he also envisions the need to create a second Internet in Space that would also enable people on Mars and Earth to connect online. Musk's Internet in Space has the initial goal of launching a series of small, less expensive satellites in near-Earth orbit, roughly 100 to 1,250 miles above Earth, which would provide Internet access to rural areas previously inaccessible to the Internet.

A rival company, One Web, has a competing vision to Musk's. Instead of a global network of some 4,000 small satellites in low Earth orbit, One Web, with financial backing from the Virgin Group, plans to build a network of 700 slightly larger satellites also in low Earth orbit. An earlier effort by Teledesic, with a similar vision and funding from Bill Gates and others, had to be abandoned as too expensive.[50]

Virtual Reality (VR)—Immersive Experiences for Education, Learning, Travel, and Entertainment in the Future—and Augmented Reality (AR)

Virtual reality or VR is a technology that is increasingly beginning to take off. It is defined as "a computer-generated environment" in which you can have a totally immersive 3D/4D (moving) experience of a different reality versus the current 2D images one sees on one's computer screen, tablet, or smartphone.

Currently a "VR headset fits around your head and over your eyes, and visually separates you from whatever space you're physically occupying. Images are fed to your eyes from two small lenses." Through VR you can *virtually experience anything*—as if you were physically there—like hiking the Grand Canyon or any beautiful spot in nature, touring the Louvre or any museum, touring another planet, experiencing a movie as if you were part of it, or "immersing yourself in a video game while still at home."[51]

<u>Progress and future projections</u>

So far virtual reality is used primarily for gaming purposes. Nonetheless, many people see VR taking off in future, with some seeing VR as the next stage beyond mobile smart phones, or even wearable tech. Various large corporations are investing heavily in VR possibilities for the future, including: Facebook, Google, and Microsoft, with Apple vying to catch up, along with Disney and Sony.

The VR vision is to imagine a future with totally immersive VR experiences impacting education, training programs in any field, such as medical surgery, entertainment, vacations (before or without going there), exhibits of museums and art galleries, house hunting from your home or office, and even meditation. Also imagine journeys with historical characters from different cultures and time periods, or journeys to space, under the oceans, or being anywhere on Earth.[52] While technological possibilities seem endless, whether capitalist or socialist, business or government, interests will limit what ends up being created remains to be seen. The more costs can come down and VR efforts can be created on grassroots levels, the more likelihood that greater creativity will occur. One concern is the effects of longterm use on the brain.[53]

Photo realism is seen as the next stage of VR development, with virtual worlds becoming so similar in detail to our regular reality that some people may choose to spend great amounts of their time in these virtual worlds, with great societal implications.

This also raises the *ultimate question* of *"what is reality" anyway?* If one can experience something, and that experience (even if virtual) is real to a given person, then is that not also "reality"? Similar questions are raised by people experiencing alternative realities through deeper states of meditation or prayer, or aided via psychoactive drugs.

Another future issue is whether VR will develop to the point that one can enter VR rooms to experience different alternative, virtual realities without needing to wear VR goggles, as portrayed in *Star Trek's* holodeck, which would definitely be less cumbersome?

Not only are scientists working on Virtual Reality, but also on Augmented Reality (AR), which is a mixture of Virtual Reality combined with or superimposed on everyday reality.[54]

3D and 4D Printing or Additive Manufacturing

3D printing allows for the use of different materials in order to print things locally and relatively inexpensively layer by layer, hence also the term "additive manufacturing." Another definition is that it "turns digital 3D models into solid objects by building them up in layers." Additive printing was first invented in the 1980s, and has been used since then for rapid prototyping (RP) of models of different possible products. In the last few years, it has begun to evolve further into a "next generation manufacturing technology" with the "potential to allow local, on-demand production of final products or parts" of products.[55]

It is already possible to use a wide range of materials in the 3D printing of various products, including "thermoplastics, thermoplastic composites, pure metals, metal alloys, ceramics and various forms of food," and it is expected that additional materials will be able to be used in future.

While 3D printing as an "end-use manufacturing technology is still in its infancy," it is believed that it could "radically transform many design, production, and logistics processes in future, especially when combined with synthetic biology and nanotechnology."[56]

One area of caution is that a 3D gun has already been printed. Given that there are already too many guns, and too much violence, in certain societies, including the U.S., the possibility of making ever more guns available to be 3D printed locally and cheaply does not bode well for creating more nonviolent societies in future.

In addition to 3D printing, 4D printing allows what is printed to move once it has been printed. It is therefore not a static product, but a dynamic, moving product.

Implications of Three New Technologies for Future Economic Systems

There are three new technologies—all discussed above—that will individually and collectively have a huge impact on future economic systems available to humanity, each briefly summarized next. These include: (*1*) *Robotics and the Increasing Automation of Production*, which is creating some new, more skilled jobs, but also eliminating many jobs, raising issues of a needed societal discussion of a guaranteed minimum income for everyone in future, with jobs for some, who could then augment their guaranteed incomes; (2) *Nanotechnology or Molecular Manufacturing*, which is raising the future possibility of being able to produce whatever is in short supply, but needed, or wanted by individuals and societies. While this could eliminate any scarcity of goods and create greater potential abundance for all, it could also have serious negative impacts on the environment—if too many goods suddenly flood and crowd an Earth already in crisis; and (3) *3D or 4D Printing or Additive Manufacturing*, which is raising the possibility of local, decentralized, inexpensive printing—on Earth or in space—of whatever is needed, also creating more potential abundance. These technologies will all have significant impacts on our human and planetary futures, whose impacts all need to be discussed not just by experts, but by humanity as a whole, so their impacts can be better understood and important priorities can also be determined.

SPACE EXPLORATION, INDUSTRIALIZATION, AND SETTLEMENT: ADAPTING TO THE TOTALLY NEW ENVIRONMENT OF THE "FINAL FRONTIER."

Humanity expanding its habitats from an Earth-only environment to combined Earth and space environments, raises many issues impacting future human evolution, as discussed next.

The Lure of Humans Going to Space Is Great

It is often remarked that humans are not biologically equipped to live, work, and survive in space without a lot of help from artificially created environments like space suits, space stations, or space colonies, or the eventual terraforming of planets like Mars.

Given all these difficulties, why does the lure of going to space have such a strong hold on the human imagination? One answer is clearly that there seems to be something in the human spirit that seeks to push back boundaries and explore the unknown, thereby helping humanity gain knowledge and evolve. Against great potential dangers, this human spirit often wins out. Humans—from ancient times onward—have also looked up at the magnificent panorama of stars and wondered what their place was in this vast universe we inhabit. All this has propelled humanity to learn more about the universe, and now to begin actually leaving Earth to explore our solar system and the universe beyond.

Great Evolutionary Leaps

People moving off their home planet into space would be making as great an evolutionary leap as when the fish first came out of the water onto the land—both adapting to totally new environments. There have been other, equally significant, earlier leaps in evolution, as when certain mammals returned from the land back to the sea to become whales, porpoises, and dolphins; when our primate ancestors came down from the trees and began walking on two legs; or when the great apes first discovered that they could use a bone as a tool or weapon—as masterly portrayed in the film: *2001: A Space Odyssey*.

Darwin's theory of biological evolution posited that when the environment changes, those members of a species most able to adapt to their new environment will have the greatest chance of surviving to procreate, which can eventually result in the emergence of new species. This principle, which Darwin called natural selection, will no doubt also prove applicable to humans who migrate to space.

Technologies Needed to Go to Space

Without computers, telecommunications, and robotics, as well as rockets and propulsion systems, no one could venture into space or communicate with space scientists back on Earth during their missions. Currently both Elon Musk of SpaceX and Jeffrey Bezos of Amazon are working on, and making real progress on, building and testing reusable rockets, which will greatly cut the costs of going to space, including to Mars. Without designs and plans for building space colonies, or international space stations, no one could ever live in space for any length of time. And without designs of space suits, no astronaut or cosmonaut could ever venture into space in a space capsule sitting on top of a fiery rocket. These are but a few of the many technologies on which any space program depends and evolves in its efforts to conquer the "final frontier" of space.

Latest Human Space Efforts, Including Mars Missions

NASA is working on a two-way Mars mission for the 2030s.[57] Elon Musk, the CEO of Space X, envisions eventually sending 80,000 people to Mars via reusable rockets he is building, which will cut the costs of such missions, though starting out with a small mission of ten colonists by 2020 or 2026—roughly ten years before NASA.[58]

Finally, Mars One is planning a private, one-way mission to Mars that began with over 200 applicants, who were then narrowed down to 100 candidates, who will finally be narrowed down (after various tests) to just 24 people (in teams of 6), who will then be paid to begin training for a Mars mission. These people are volunteering for a one-way mission to never return to Earth, showing a real adventurous spirit and willingness to truly explore the unknown.[59]

NASA, as a government agency, can only send people on a two-way mission, which requires twice the fuel and a return capability, making it a much more complex venture. Some people question whether the Mars One mission will ever actually occur, due to technological and funding issues.[60]

Longer term Human Futures In Space

Other very significant problems facing humans who live in space for extended periods include bodies becoming elongated and bones weakened due to the absence of gravity. This makes regular exercise essential and can even make it difficult for humans to adjust to life on Earth again after they have been in space for too long. Eventually, when permanent space colonies exist, and children start being born in space, their natural environment will be space and they may also find the heavier gravity of Earth difficult to adapt to.

Other future issues for people living in space colonies include the possibility that space colonies could eventually declare their independence from the governments or private space programs back on Earth that originally sent them to space—not unlike the American colonies that finally declared their independence from their colonial mother country,

England. And if intelligent extraterrestrial (ET) life and alien species are encountered in space, this will certainly have a profound impact on future human evolution both in space and on Earth.

The above challenges for humans living permanently in space in the future could well lead over time to the emergence of a new space-faring posthuman species that differentiates itself increasingly from the humans left back on Earth.

While a very small percentage of humanity will ever go to space, it is important to remember that 99.99% of humanity will remain here on Earth, necessitating that humans learn to better co-habit this planet together, as well as learn to become better caretakers of Earth—the only home most of us will ever know.[61] Due to concerns that climate change or armed conflict and warfare could undermine or destroy our human efforts and presence here on Earth, some people (such as Stephen Hawking and Elon Musk) argue for the necessity of humans going to space, to ensure that at least some members of our human species survive and carry on the human experiment. For many others, space remains the "final frontier" that beckons humans for numerous positive reasons.

CONCLUSIONS

Part II has briefly reviewed some of the key developments and issues raised by eleven different technologies that will all be impacting our future human evolution and lives in important ways, including impacting each other. These eleven technologies collectively offer a good introduction to some of the key issues facing humanity as we move into the future, including the all-important question of whether what it means to be a human being may also be changing in important ways.

More dialogue between proponents of these different technological views is also essential—which should include not only experts in each area, but the general public, who will be impacted by developments in all these areas. Too often experts stay in their own areas of specialization, but the issues raised here are so important that both experts and the general public need to join in in-depth discussions about how present day policies and decisions are impacting the future of human evolution and human nature as we know it.

Beyond the technological issues covered here, additional issues that will also profoundly impact future human evolution include Earth sustainability, civilizational diversity and interdependence, evolving consciousness-based perspectives, and (as an extension of space exploration perspectives covered here under technology) whether extraterrestrial life, including intelligent life, will be found to exist in our solar system, Milky Way Galaxy, and beyond, which seems increasingly likely. All these issues will be discussed in Part II of this author's forthcoming book *Options for Future Human Evolution: Cutting-Edge Issues for Humanity and the World*.

Though not covered in detail here, an important evolution of human consciousness—to dynamic, interdependent, complex, whole systems thinking and worldviews, in line with the thinking and process philosophy of Alfred North Whitehead—would help considerably in providing the framework needed for more consciously, responsibly, and creatively looking at, and responding to, all the many issues confronting humanity and the world today. These issues include how humans respond in an interdependent, responsible way, not only to the many new technologies impacting our lives, but also to the many crises currently facing the Earth and other endangered species, as well as how to build bridges of understanding between all the diversity of the world's peoples and their different cultures, civilizations, and religions, which are increasingly interacting with each other within our increasingly interdependent world. In this way, the ideas of Alfred North Whitehead remain as prescient and relevant as when he first shared them.[62] It is noteworthy that futurists and evolutionary thinkers also tend to have a dynamic, interdependent, complex, whole systems worldview and framework, which they use for analyzing the many changes and evolution occurring in our world and lives. This provides important parallels between Whitehead and the fields of Futures Studies and Evolutionary Studies—two fields of great interest to this writer.

SEVEN

Regarding Humanism:

Some Observations Concerning the Tibetan Buddhist and Transhumanist Dialogue

Sean K. MacCracken

THIS CHAPTER[1] IS WRITTEN with a twofold aim in view. To existing dialogues between Buddhists and futurists it contributes: 1.) the observation that meaningful dialogue can take place between science and religion to the extent that each can bracket its respective dogmas and agree upon the sort of shared epistemic categories that render fruitful exchange of ideas possible; and 2.) the argument that transhumanist philosophy thus far errs in unconsciously inheriting transcendentalist (anti-material, anti-embodied) presuppositions. And, while the Nāgārjunian philosophy favored by Tibetan Buddhists is only one of many possible philosophies to mitigate such transcendentalist tendencies, Nāgārjuna's Madhyamaka doctrine is especially well-suited due to the overlap in vocabulary it shares with the proto-transhumanist philosopher, Derek Parfit. This chapter contains a review of Parfit's philosophy with respect to personhood; a consideration of Parfit's affinity with Buddhism; and some suggestions concerning future avenues of research, and what a futurism might look like that bears more of a Buddhist inflection, and less of a transcendentalist inclination.

THE TRANSHUMANIST LEGACY

Transhumanism, broadly conceived,[2] has come a long way between the utopian techno-futurism of the 1990s and the current global climate, however our present moment is to be remembered. Among many recent achievements celebrated by transhumanists, one prime example is the appointment in 2012 of the celebrated inventor, accelerationist, and renowned prognosticator Ray Kurzweil to a full-time position at Google, where he works on machine learning and language processing. Another is Aubrey de Grey's SENS Research Foundation (Strategies for Engineered Negligible Senescence), founded in 2009. Far from having been consigned to the cultural recycling bin along with Seattle grunge bands and other 90s artifacts, transhumanism remains very much alive and active. It is no longer—if it ever was—a fringe movement, but is a constellation of ideas influencing not only research and development, but conversations concerning philosophy and ethics in Silicon Valley and around the globe. And at the same time, it should be noted that there is a transhumanist crisis of outreach. On the one hand, there is no doubt that transhumanist discourse has an audience, sparking the interest of populations intrigued by its promises of radical life extension and other presumed improvements to the human condition. On the other hand, one recent survey study has provided substantial evidence concerning demographic imbalances amongst transhumanist enthusiasts. In this survey study, respondents were overwhelmingly male and white—even relative to the already skewed standards of science, technology, engineering, and mathematics (STEM). Such imbalance rightly suggests to transhumanists a need to widen a program of outreach and human concern, an endeavor in service of which a number of different approaches are possible.[3] On the political side, a worldwide collection of individuals and organizations have become co-signators to the *Technoprogressive Declaration*, which seeks to advance transhumanist aims from the position of a socially democratic ethic, in counterpoint to what is perceived as the more heavily funded right wing of transhumanism, developing along libertarian lines.[4]

Both these developments—transhumanism's demographic self-assessment and the emergent technoprogressive turn—are salutary developments to be applauded by socially progressive futurist communities. At the same time, there remains some question as to how such developments cohere. And there remains the important question—what, if anything, might be difficult or objectionable in the intellectual assumptions at the root of transhumanism thus far conceived? The present chapter tackles this crisis of outreach from a philosophical perspective. It seeks to communicate faithfully and to critique the prevailing "hard transhumanist" philosophical rationale behind the category of the person. To explain, hard transhumanism is here understood as a transhumanism of a stridently functionalist-reductionist philosophical tone. This is a tone that—as much as any of its political manifestations—appears to render such transhumanist discourse distasteful to other futurists—reductionists and non-reductionists alike. In addressing what many (non-transhumanist) futurists may find particularly hard to swallow in the influential early articulations of transhumanist philosophy, this chapter seeks to define a wider spectrum of space for thinking about the category of the person—a space that opens up transhumanist discourse to greater intellectual and emotional (and in the process, perhaps gender and ethnic), diversity. Thus this chapter seeks to contribute for philosophy something analogous to what the Technoprogressive Declaration accomplishes for politics. The first goal of this chapter, then, is simply to furnish an introduction to transhumanist philosophy as it is thus far deployed.

A secondary consideration is the still-nascent encounter between transhumanism and Buddhism. Although religion is by no means a necessary environment for the cultivation of ethics and morals, world religions in their most benign expressions do not simply hold humanist values in common with a technoprogressive agenda. They also provide certain resources that—though they may be subject to rational inquiry, should not be dismissed out of hand, in that they are venerable repositories of human concern. Buddhist epistemology, in particular, represents a large body of experimental psychological data concerning the elusive nature of the category of the person.[5] Such Buddhist epistemology is a very old technology that awaits cross-disciplinary interaction with emergent technologies. Buddhism's often-celebrated project of heuristic method (*upāya*) and empirical observation lend the religion a certain natural affinity to contemporary scientific observation.[6] Thus, if meaningful cooperation between transhumanism and religious groups is to take place, Buddhism need not be the only religion, though it is certainly well-positioned to be among the first and foremost.

The foregoing observation—concerning the encounter between transhumanism and Buddhism—is not merely theoretical, but is intended to highlight an already developing dialogue between Buddhism—especially Tibetan Buddhism—and transhumanism.[7] From very early on, Buddhist philosophers have been concerned with mereological—that is, "parts-and-wholes" analyses of the category of the person, founded upon reductionist methods. Such mereological analysis stems from the observation that—for later Buddhist authors—the Atomistic Reductionism of the early Abhidharma does not represent the subtlety of the Buddha's observations with sufficient fidelity—giving as it does an account of reality reduced to the purportedly essential constituent atoms of experience (*dharmatās*), and thus inadequately accounting for the Buddha's cardinal observation concerning the non-essentialized (*an-ātman*) nature of conditioned phenomena. Within the Indian legacy of Buddhist philosophy, such mereological inquiry has prompted a spectrum of outlooks from absolute-idealism to epistemic-idealism (as in the Vijñānavāda textual tradition),[8] or else a spectrum of subtle methods of what might in the twenty-first century be called a type of deconstruction (as in the Madhyamaka textual tradition): those ranging from emphasis on Reductio Arguments (Prāsaṅgika) to Autonomous Reasoning (Svātantrika).[9] Tibetan Buddhist philosophy generally embraces the Madhyamaka as the more subtle philosophy, with the Vijñānavāda in a secondary position. In so doing, Tibetan Buddhist discourse arguably inherits from India a philosophical tradition that is increasingly less structuralist, and more process-oriented.[10] Many further bridges of understanding between transhumanism and Buddhism remain to be forged. Intended as a philosophical overview and statement of the field, this chapter is but a small step in that direction. Specific to the topic under consideration, transhumanist philosophy famously applies mereological analysis to the category of the person. Buddhist philosophy undertakes a similar project, over a span of many centuries. Such is the basis for direct comparison.

UNDERSTANDING ONE ANOTHER: EPISTEMIC CATEGORIES

In the Indian philosophical tradition, formal debate between one school and another traditionally requires that interlocutors agree on which epistemic categories they hold in common before formulating and comparing arguments. Such agreement renders the discussion in an appropriately free transdisciplinary space. For example, direct perception (*pratyakṣa*) must at minimum be admitted as a valid epistemic category. Likewise, to rely upon inference (*anumāna*) may be agreed upon as permissible. These and other categories, interrelated and perhaps further qualified according to the doctrines of a particular school, make meaningful debate possible without a tradition adverting to the "word" (*śabda*) of canonical authority. For example, the authority of the Buddha's word is not something a Hindu, Jaina, or Hedonist (Cārvāka) philosopher is obliged to regard as admissible evidence. Yet, such testimony might be optionally considered if the Buddha or his commentators appeal sufficiently to logical reason, including principles of noncontradiction, and so forth. Likewise, Buddhists and non-Buddhists alike since the Buddhist epistemologist Dignāga have attended to the distinction between an "inference for oneself" (*svārthānumāna*), for which one's own preferred epistemic categories are sufficient, vs. an "inference for others" (*parārthānumāna*), which intends persuasion via sensitivity toward the epistemic warrants of another person, philosophy, or textual tradition. The appeal to canonical authority is never a sufficient basis for constructing a sound argument. This is all worth keeping in mind by way of rough analogy to any contemporary dialogue between the separate spheres of science, philosophy, and religion. To paraphrase an idea put forward by the Tibetanist (and ordained monk) Robert Thurman, Western science and the specialized "inner science" component of Indo-Tibetan religions have something to offer one another only to the extent that each can put aside its respective dogmas and mutually define a common vocabulary emphasizing thesis, data, analysis, and falsification of data.[11] Outside of such agreements, vague harmony of purpose and fuzzy ethical outlooks may be shared in common, while deeper accord of purpose, sharper identity of understanding, and assiduous coordination of effort remains forever elusive. To the extent that both partners in dialogue confine their interlocution to direct perception, Buddhism and transhumanism enjoy a shared space of discourse. The use of inference, perhaps, is debatable. In such common space of discourse, one must allow that the statements of the Buddha and his countless commentators cannot be taken on faith, but must await systematic testing and are subject to falsification. On the other hand, despite the modern ascendency of scientific method with its attendant salutary developments, blind scientistic dogma is likewise to be avoided. What remains unfalsifiable by current scientific inquiry, then, is not something upon which science is entitled to issue dogmatic rulings. Science deals primarily in data and facts and reasoned conjectures, not in rigid beliefs and dogmatic convictions. For example, *a priori* beliefs concerning the absolute materialist nature of reality are not to be confused with data and conclusions on the same.[12] It is the intention of this chapter to operate within a transdisciplinary space of analysis, in hopes that philosophers and scientists alike can partake in this playground of thought, without either side grabbing all of its marbles and stomping off in anger.

SECTION 1: DR. MAX MORE AND THE PARFITIAN LINEAGE

The "hard transhumanist" philosophical lineage this chapter wishes to analyze and critique is strongly *Parfitian* in nature. One of the most revolutionary, controversial, and important philosophers of the twentieth and twenty-first centuries, the recently departed Derek Parfit (1942–2017) devotes substantial effort in his landmark publication *Reasons and Persons* (first published in 1984) to demonstrating that commonsense notions of personhood are incorrect. An appropriately reductionist analysis instead leads the reader to the unsettling, counterintuitive, yet eminently rational (and thus purportedly correct) conclusion that persons are nothing more than their physical and psychological constituents. To account for this totality is to be in possession of all the facts, however such facts are then interpreted. Persons cannot be reasonably demonstrated as some "further fact" beyond this physical and psychological constituency. Moreover, in the hypothetical event that a person's physiology and psychology (disposition, talents, and memories) could somehow be replicated, the resulting copy would be a person "just as good as" the original. And, in the event of the original's demise, such a copy should rationally be considered to *be* the original person, on what Parfit terms the "closest-continuer" theory. That is, the causally related copy that bears the closest resemblance to the original *should be considered as* the original. For it is not an identity relation that matters to Parfit, but what he terms *Relation R* or *R-relatedness*. In this case, the *R* simply designates a sequence of persons-over-time. What I conventionally take to be myself concerns how I stand causally in relation to a sequence of diachronic *selves*. Without the circular logic of *a priori* presupposing personhood as either a Cartesian ego or else a "further fact" over and above the constituent parts of the person, persons over time may be reductively understood as that which proceeds in physiological and psychological connectedness and continuity from the present person, *given the right kind of cause.* It is this latter criterion—the "right" kind of cause—that gives *Relation R* a wide range of possible interpretations within Parfit's scheme. With present day technology, the only cause of *Relation R* that is both reliable and normal (that is, conventional) is the persistence of brain and body, whereas in the future we may speculate that brain-body transplants or partial brain replacements could furnish a reliable cause of *Relation R* that breaks from, and thus challenges, what is considered the conventionally normal range of causes of *Relation R*. In other words, our currently rather tautologous notion of persons-over-time as persons with the same body and brain may stand in need of revision as time and technology progresses. Only *you* are you today, but in the course of the future it may come to be the case that other brains or bodies can and should also be considered you, if any of your duplicates survive the original you. Such relationship between person-states, or persons-over-time, is said to be *R-relatedness*. Parfit allows that *Relation R* will enjoy a wide spectrum of possible interpretations. In some interpretations, *Relation R* need not have a conventional cause at all, but simply a reliable one.

Parfit's analysis is a significant precursor to transhumanist thought, or, as will be argued here, a "classical transhumanism," though this is an etic designation. Parfit does not self-identify as a transhumanist, though he is cited as though he were one.[13] Like many transhumanists, however, he undertakes a project of what he describes as revisionist, as opposed to descriptive, philosophy.[14] A self-avowed transhumanist, Max More follows form in his dissertation, making normative claims with respect to how personhood ought rationally to be viewed. Parfit's discussion of persons is couched also in ethical discussions (the *Reasons* of his title). More, though evincing his lifelong concern with cryonics and medical ethics, remains primarily concerned with the category of the person, and not ethics. It is this common thread of reductionisms with which the present chapter is concerned.

Dr. Max More is arguably the most mature and influential living exemplar of a transhumanist *philosophy*. His 1995 dissertation presents itself as an extension and clarification of Parfit's *Reasons and Persons*, while making divergent normative claims of its own.[15] Significantly for the transhumanist outlook, More avowedly parts ways with Parfit in More's emphasis upon continuity over connectedness, in contrast to Parfit's purported emphasis on connectedness over continuity. Such terms invite a bit of explanation as to their Parfitian meanings. Identity is a transitive relation. What replaces and stands in for identity (*Relation R*) must, like identity, be transitive. Something like *psychological connectedness*, "the holding of particular direct psychological connections," is not transitive, as persons over time normally resemble themselves closely from day to day, but it does not follow that they do so from decade to decade. What is transitive, for Parfit, is *psychological continuity*, "the overlapping of chains of *strong* connectedness," where *strong* connectedness concerns the number of "direct connections" between person-phases being maintained at "at least half."[16] In terms of the difference between Parfit and More, it remains important to More that persons have a causal relation to their persons-over-time. Yet, it is less important that such persons remain self-similar over time, quite the contrary. Such a stance is significant given the transhumanist emphasis on change, evolution, and the so-called "morphological freedom" of the individual.[17] To More, it should be the right of every human being to pursue positive and radical changes to one's condition and character, according to their personal volition and evolving self-understanding. So connectedness to More is desirable to an extent, but only continuity is personhood's *sine qua non*.[18] Already present in Parfit's philosophy, then, is the sort of thinking that would license partial—or full—brain and body replacement, with the insistence that personhood is retained. Already present in Parfit is the foundation of a philosophy that legitimizes transhumanist ideals such as so-called "mind uploading" in functionally identical bodies.[19] With Parfit, the argument is that personhood should be recognized based on *R-Relation*, which may be variously interpreted. With More, for his part, the argument is that *Relation R* is to be correctly recognized as based upon any reliable and intentional cause, but that causes considered "normal" can and should change over time with the progression of technological and cultural norms. Additionally, More's emphasis on continuity over connectedness opens the doors onto radical changes to persons and the definition of personhood, including "mind uploading" in new bodies that change and depart significantly from the original body.

Unzipping the file: the Parfitian-Morean vision unpacked

Parfit's *Reasons and Persons* is not something one reads lightly, even if focused only on its "Persons" aspect. Like the proverbial zip file, having overviewed the Parfit-More textual tradition, some unzipping of the zip file is now warranted. To elaborate, for this textual tradition, R-Relatedness deposes the commonsense (but purportedly incorrect) notion of personal identity. The reader wishing to think along with Parfit is referred to Part Three of *Reasons and Persons* to trace the development of this idea.[20] For the purpose of this chapter, no more than a summary of R-Relation is warranted. Parfit appears to understand *R-Relation* as departing from the reductionist standards of physicalist-reductionists (I am essentially my body) and neuropsychological-reductionists (I am essentially my brain), while building upon the two. The fact of *Relation R* is necessary and sufficient for survival, regardless of how one understands the cause.[21] Parfit understands the psychological criterion for survival as follows:

> *The Psychological Criterion:* (1) There is psychological continuity if and only if there are overlapping chains of strong connectedness. X today is one and the same person as Y at some past time if and only if (2) X is psychologically continuous with Y; (3) this continuity has the right kind of cause; and (4) it has not taken a 'branching' form. (5) Personal identity over time just consists in the holding of facts like (2) to (4).[22]

Again, recall that "the right kind of cause" is a deliberately ambiguous phrase for Parfit, opening this criterion to debate and interpretation. To explain the term "branching"—Parfit will elaborate upon the "branching" form of a self in some of his thought experiments in imaginary scenarios that—were they to be carried out—would certainly offend currently established medical ethics. Though, perhaps, there are no thought-crimes! For example, the bisecting and transplantation of one person's brain into two bodies, on the theory that a person can survive with only a portion of their previous brain function, and so could survive in two separate bodies each with half the brain. Such a practice would be, of course, without current justification under the Hippocratic Oath, and is viable only theoretically due to the present impossibility of bisecting a living brain stem.

The remaining above psychological criteria are fairly straightforward, with the exception of the deliberately ambiguous "right kind of cause." Acceptance of the Psychological Criterion is of three types. Parfit explains, "These differ over the question of what is the right kind of cause. On (1) the Narrow type, this must be the normal or conventional cause. On (2) the Wide type, this could be any reliable cause. On (3) the Widest type, the cause could be any cause."[23] Though Parfit does not use this terminology, causes could essentially be (1) conventional, (2) unconventional, or (3) miraculous (by current standards). Such designations make more sense when applied to the examples to follow. What is conventional, of course, should be understood to change over time according to social, legal, and technological norms. This is an understanding that transhumanists will be keen to exploit, arguing that social and legal convention is unfairly constrained by what is technologically possible, and by society's imagination of what will become possible. What is reliable, of course, is also subject to subjective interpretation, as the means of "reproducing" persons will surely develop incrementally in reliability, and will not arrive all at once as fully realized and 100% reliable means.

The physical criterion is understood as follows:

> On the Physical Criterion, a person continues to exist if and only if (*a*) there continues to exist *enough* of this person's brain so that it remains the brain of a living person, and (*b*) there has been no branching in this physical continuity. (*a*) and (*b*) are claimed to be the necessary and sufficient conditions for this person's identity, or continued existence, over time. On the Narrow Psychological Criterion, (*a*) is necessary, but not sufficient. A person continues to exist if and only if (*c*) there is psychological continuity, (*d*) this continuity has its normal cause, and (*e*) it has not taken a branching form. (*a*) is required as part of the normal cause of psychological continuity.[24]

Taking up Parfit's penchant for thought experiments, More supplies a number of science-fiction- (or, as some hard sci-fi fans might have it, fantasy-) inspired examples of personal survival along with designating acronyms. These can be briefly summarized as follows:

1. (MBT) My Brain Transplant—The brain is transplanted into "a decerebrated clone."
2. (TT) The Transporter—The brain and body are scanned and disintegrated. Of different molecules (but the same elements), the body is rebuilt in a different location based upon the scanned pattern.
3. (AHR) A Heroic Reconstruction—Stunningly advanced technology permits the intentional recreation of specific persons of special historical interest down to the quantum level, based on detailed historical data.
4. (OPR) Omega Point Resurrection—A sufficiently advanced society "in control of all matter and energy in the universe" recreates every possible living person down to their quantum states in such a way that incidentally reproduces a specific person.[25]
5. (LPU) The Luckiest Person in the Universe—A person matching the exact description of another historical person is spontaneously generated as the extremely rare "outcome of a completely random and acausal quantum process."[26]

Such far-flung examples alone might give certain readers a bit of a nosebleed. These theoretical scenarios are grounded, if that is right word, by considering the way *Relation R* applies to each. Following Parfit's own sentiment concerning thought experiments, it is not the literal, future viability of such examples that is important so much as what they serve to illustrate about the reader's own instinctive convictions concerning personhood and survival.[27] Note that all five of the above examples presuppose psychological connectedness and continuity. None satisfies a contemporary cause for survival, as none is currently possible. Only (1) MBT satisfies the physicalist criteria of the Narrow Reductionist (NR) view—that one is essentially one's brain. (2) TT to the physicalist or NR view is tantamount to death as it destroys original body and brain, but is not so for *patternist* philosophers, who hold *Relation R* based views from the Wide Reductionist (WR) view and beyond—that is, views that permit not just a conventional cause but any reliable cause.[28] (3) AHR conforms to what More has called a conservative interpretation of the wide reductionist view (CIWR), which requires only some sort of causal relation, regardless of its reliability.[29] In (3) AHR, the person chosen for reconstruction is selected by scientists specifically based upon their perceived usefulness to society, which sets up a causal relation of intentionality. (4) OPR and (5) LPU correspond to what More calls the extreme interpretation of the wide reductionist view (EIWR), but in its moderate (M-EIWR) and radical (R-EIWR) interpretations respectively. (4) OPR admits of survival on the basis that there is some causal relationship (in this case, intentional re-creation of the person), despite the lack of direct connection to the earlier person (as in this case, all persons in history are re-created without discrimination), and is thus M-EIWR. (5) R-EIWR would countenance survival in the case of LPU, as this view requires no causal relationship at all between the original person and the duplicate, only qualitative similarity, and is thus R-EIWR. These categories may be tabulated according to Table 1 (next page).

More, following Parfit, has thus demonstrated a continuum of scenarios with respect to personal survival in which he himself confidently argues in favor of personal identity being maintained through as far as what turns out to be the middle of five views—a "conservative" interpretation of Wide Reductionism, on the basis that a direct causal relationship should be the defining criteria for survival of personhood. He rejects the latter two scenarios on the basis of the absence of such a direct causal relationship.[30] Such apparent moderation or even conservativism is crucial in More's presentation of his own philosophy. Because of his emphasis on

The view:	Admits survival in	Wherein there survives
Narrow Reductionist	My Brain Transplant	Brain only, through reliable cause
Wide Reductionist	Teletransporter	Relation R, through reliable cause
Conservative Interpretation of Wide Reductionist	A Heroic Reconstruction	Relation R, through direct intentional causation
Extreme Interpretation of Wide Reductionist: Moderate	Omega Point Reconstruction	Relation R, through indirect intentional causation
Extreme Intepretation of Wide Reductionist: Radical	Luckiest Person in the Universe	Relation R, acausally

Table 1

morphological freedom and personal volition, the binding condition of a direct causal relationship has the effect of constraining personal survival significantly. With More's emphasis on personal growth and transformation, including radical modification of the self and body, an emphasis on causality serves to establish a clear limit on what might otherwise seem a potentially totally unbounded idea of personhood. Personal will or volition is thus for More both a liberating and a binding condition—persons over time are free to modify themselves, yet it is precisely direct intentional causality that defines what personhood-over-time means, not indirect causality or mere exact resemblance.

How personhood is defined for Parfit, then, concerns *Relation R* in place of a conventional notion of identity. The person generated by the output end of the Teletransporter is the same as the person disintegrated by the input end. How personhood is defined for More extends as far as any reliable and direct causal relationship. The person reconstructed centuries later based on historical data is the same as the person who died a natural death centuries previous.

The Present Moment: How It All Matters in The Early Age of the Singularity

The present chapter is thus far concerned with description. Before proceeding to analysis of the philosophy so far described, it seems sensible to do the work of relating More's observations across disciplines to present-day scientific achievement. At the time of this writing, some two decades after More's dissertation, it would seem likely that real-life personhood controversy will likely arrive soonest with strong artificial intelligence (Strong AI). That is, science will progressively develop AI systems that closely mimic wholes or parts of human brains based on scanning techniques (progressively satisfying Parfit's *Relation R* criteria for psychological personhood) prior to either 1.) artificial human bodies, 2.) brain transplants, or 3.) the coding of Strong AI systems line-by-line, without assistance from such scanning techniques.[31] While further and more elaborate speculation is outside the discipline and purview of this chapter, this basic assumption is worth noting for the following reasons. First, and most obvious, this constrains the horizon of what is imminently possible to a discussion of disembodied strong AI personhood (which might nonetheless satisfy *Relation R* criteria in terms of Parfit's neuropsychological criterion, and physical criteria to the extent that AI persons are given autonomous freedom to act in the world). Second, and only slightly less obvious: given the present range of possibilities as opposed to fantasy scenarios, this places More not in the conservative middle-space of a continuum in accounting for personhood, but at its outer extreme end! More's philosophy would appear to grant some personhood status to strong AI, and even "closest continuer" status to strong AI that survives the original person from whom it was copied. Third, then, is the conclusion that More's first three thought experiment scenarios—MBT, TT, AHR—are of particular interest (being closer to speculative fiction than technology-as-magic fantasy), though they may be of immediate interest only through modification. To explain, brain transplants remain impractical while modern science is beginning to understand how a person's brain may be scanned and mimicked without damage to the original brain. More's above examples break down somewhat when one reflects that a scenario more immediately likely within the next generation is the controversy concerning a living organic person being an older contemporary of his or her (not necessarily bodied) AI duplicate, a duplicate that may change and "evolve" at a radically different pace, and which may go on to far-outlive the organic duplicate from which it (or should we say he or she?) was generated. In essence, much current reflection attends questions roughly on a continuum between the TT example (what personhood inheres in a *Relation R* duplicate that is contemporary with its original, and that survives its original?) and the AHR example (what personhood inheres in a *Relation R* duplicate produced from a recently or long-deceased original?). Future research in STEM fields will no doubt address itself to this question.[32] It is mentioned here only as a basic assumption guiding the writing of this chapter.

To summarize the description thus far, Parfit's *analysis* is one of conscious reductionism. It is in a significant sense structuralist and atomistic—Parfit would say *deterministic*. Parfit's thought experiments are intentionally guided by a determinism with respect to persons. In the Teletransporter case, which is originally Parfit's, the reader is asked to imagine that the duplicate person either *is* or *is not* the same person as the original. That "both" are the same or that "neither" are the same person are not acceptable answers. Such determinism is geared toward testing the reader's convictions with respect to the nature of personhood. Parfit's philosophy is also reductionistic in the sense of not presupposing a "further fact" such as a Cartesian ego. That is, he rejects non-reductionists who hold personhood to inhere in a separate self, such as an immortal soul, but also those claiming that personal identity is a distinct condition that is not a separate self.[33] In not succumbing to such a presupposition, when we know that a person emerges from the Teletransporter with an intact *Relation R* to the original person who entered the Teletransporter, we are in possession of all the facts. Interpretation of those facts, for Parfit, is secondary. Moreover, because persons-over-time with a reliable, conventionally normal cause (e.g., the simple progression of time and age) are *functionally* indistinguishable from persons-over-time with a reliable, unconventional cause (e.g., the Teletransporter), they are "just as good" as one another. It is not identity that matters, but *Relation R*, consisting of psychological connectedness, and psychological continuity as overlapping chains of strong connectedness. The irruptive event of being disintegrated by a Teletransporter is only a more dramatic version of the change and discontinuity one experiences from day to day. This for Parfit is a salutary realization, diminishing his self-interested concern. It is "liberating and consoling" for Parfit to be convinced of the truth of his claims.[34] However, it may not require any academic arguments to demonstrate how these very same conclusions might be viscerally, instinctively, upsetting for immanentalist or physicalist people of various kinds—anyone from certain scientific materialists to Spinozists, and many categories in-between. For such persons, the fact of the duplicate person emerging from the Teletransporter is little comfort to the person about to be disintegrated, however useful to society such a duplicate is, or however charming!

More, for his part, is able to countenance such reductionism without any difficulty. Some two decades after his work on Parfit, in a contribution to *The Transhumanist Reader,* More has—as an important consultant on precisely how transhumanism is academically defined[35]—taken it upon himself to speak on behalf of transhumanists with the confident assertion that most of their diverse philosophies share a common commitment to materialism, physicalism, or specifically functionalism, notably revisionary and eliminative materialisms. The outcome of such commitment is that, for the majority of transhumanists, persons must be instantiated on some sort of medium, though such a medium need not be human or even biological. Transhumanist philosophy, for More, comprises a spectrum of functionalist outlooks. There seems little in this late articulation of More's thought to suggest that he has much departed from his dissertation work. The functionalism to which More subscribes, it may be assumed, is a Parfitian-inspired functionalism in which *R relatedness*, and not identity, exhaustively accounts for criteria of personhood. However, in this later work, More frames his assessment of transhumanist outlooks less in terms of persons and more in terms of mental states. Here, the extreme transhumanist formulation is presented as a Churchland-style Eliminative Materialism. That is, "some of our common psychological concepts such as belief, desire, or intention are so poorly defined that they will be found to lack any coherent neurological basis." As eliminativism, broadly speaking, is a central theme of the present chapter, it is worth clarifying that Churchland is thus eliminativist with respect to common-sense or "folk psychology" mind states, not necessarily with respect to personhood per se. In what should by now be a rhetorical strategy familiar to the reader, More contrasts against this extreme formulation a less radical transhumanist spectrum of outlooks. "Revisionary materialism" he tells us, "takes the intermediate position that mental states may be reducible to physical phenomena, but only after some significant changes are made to the folk psychological concept."[36] That is, transhumanist mainstream discourse—far from wishing to eliminate notions of belief, etc.—wishes a revisionist program in accounting for concepts like belief in terms of what is materially demonstrable within neurology. Transhumanist extremists don't countenance neurologically ill-formed concepts like belief. Transhumanist moderates wish to revise concepts like belief, with ever-greater fidelity to what can be materially, neurologically demonstrated to be true. More's revisionism is something to which this chapter responds in its conclusion.

What is significant in More's presentation is a situation about which he is—to his credit—more or less explicit. That is, transhumanism has undertaken certain endeavors around some of the most basic questions of human experience—personhood, but also life and death. Such questions have involved transhumanists not only in philosophy but in metaphysical speculation. In some sense, to paraphrase More, transhumanism often occupies a cultural "niche" that overlaps significantly with that of religion.[37] Departing from More, it is a basic conviction of this chapter, however, that transhumanism is at its least successful when it seeks to advance its own tenets of faith—thus departing from its self-avowed domain of scientific inquiry, while attempting to co-opt for itself what is properly in the domain of metaphysics and even religion. In brief, the younger movement of transhumanism embarrasses itself by trying to prematurely seize control of *the religious* from much older institutions, when it might better serve greater collective interest by instead humbly advancing scientific method either in distinction from or in partnership with religion. Transhumanism, it is here theorized, alienates would-be collaborators because it too often uncomfortably and awkwardly appears predisposed to depose, disparage, and supplant religious tradition and substitute an eschatology of its own around the promise of life-extension technologies—millenarian, utopian, teleological, and overwhelmingly faith-driven. Such technologies, moreover, manage to be further disembodied to the extent that they are functionalist. Thus the transhumanist eschatology does not overcome but rather enshrines and fortifies a Western preoccupation with proverbially "up-and-out" varieties of transcendentalism—there is a sort of disciplinary imprecision, and thus blurring between the salutary goal of overcoming human limitation and the inveterate Western transcendentalist goal of rejecting the biological and disparaging the embodied. Such a rather sophomoric orientation belies the reality that transhumanism is now, in reality, a mature movement relative to its status two decades ago, and is hereby called upon to ease into the role of collaboration and dialogue, rather than the role of radical, rebel, and agitator.

None of this is intended to condemn the philosophy of Dr. More specifically. Nor is acknowledgment of his awareness of the situation intended to damn with faint praise. On the contrary, the hope is that philosophical engagement with transhumanist ideals can more easily cross boundaries of discipline where metaphysical speculation cannot. It is for this reason that More's writing was chosen as opposed to say, that of Ray Kurzweil—where brilliant engineering often sits awkwardly alongside millenarian fervor and breezy pop cultural humor, with little of scholastic or philosophically persuasive interest to systematically win over any skeptic. Kurzweil seeks not to logically persuade so much as compel the reader or listener by the seeming inevitability of the comforts and advantages he promises technology will provide. For all of More's own achievement, however, it is nonetheless telling that More characterizes transhumanists in terms of what often sound like faith convictions, rather than strictly logical or systematic inquiry, asserting about most transhumanists, for example:

> They believe that our thinking, feeling selves are essentially physical processes. While a few transhumanists believe that the self is tied to the current, human physical form, most accept some form of functionalism, meaning that the self has to be instantiated in some physical medium but not necessarily one that is biologically human—or biological at all.[38]

Here, perhaps, the transhumanist "mainstream" might take a cue from its eliminativist materialist "extremists" and ask, is the aforementioned *belief* even a well-formed concept? Where does it come from?

Moreover, it may go without saying that being a futurist in general, or an enthusiast for transhumanist projects like Aubrey De Grey's life extension project, need in no way presuppose functionalism, eliminativist materialism, or even reductionism! But what are the practical philosophical outcomes of More's Parfitian project of reductionism, and how might such outcomes be situated with respect to non-reductionist philosophies? One possible answer comes by looking not at More, but returning to Parfit.

SECTION 2: PARFIT'S ACHIEVEMENT: THE EMPTY QUESTION

Parfit's philosophy accomplishes a fascinating thing. It insists upon deterministic *analysis*, by means of which it arrives at an unproblematically nondeterministic *resolution*. That is to say, taking the Teletransporter thought experiment as an example, the reader is at first asked to determine whether or not the original and duplicate persons are the same. However, Parfit marshals evidence against the sensibility of such a determination, seeking to undermine the logic behind it and to demonstrate its status as what he calls an *empty question*. If one is in possession of certain facts—the percentage of cells from the original person inhering in the duplicate, or the number of qualitative traits assigned as a quantitative percentage of the original (e.g., more or less than half of his or her creativity, etc.)—then one is simply in possession of all the facts and may interpret them variously. To assign a threshold of fifty percent sameness, for example, as a criterion of identity, remains arbitrary. But the same is also true for whatever percentage one determines. For Parfit, in other words, *Relation R* stands in for identity because the former is based upon demonstrable causal relationships (facts), while the latter is entirely a matter of one's subjective determination (interpretations). *Identity is all interpretation, but* Relation R *is intended to be all fact.*

At this point, some trenchant process philosophical critiques are possible in which one could attempt to deconstruct the assigning of numerical values to qualities such as creativity.[39] Such critiques would not contradict the central point of Parfit's conclusions, however. Parfit means only to consign questions of identity and persons-over-time to a domain of interpretation. Such questions are *empty*. That is, they are various interpretations of the same facts, rather than multiple answers, one of which is correct. *Are the original user of the Teletransporter and her duplicate the same person?* This is one such empty question. One cannot answer it directly, but can advert to ways of demonstrating a *Relation R*. As has been shown, *Relation R* is intended to be more concerned with facts than interpretations and is, at root, concerned with a spectrum of causal relationships. While Parfit provides this overarching spectrum of theories, More forcefully argues in favor of what he has called the Conservative Interpretation of Wide Reductionist (CIWR) *Relation R* theory, which admits of any and all direct reliable causes, but dismisses any apparent continuity of identity between persons that is an apparent continuity either without direct cause, or without any cause at all. It is from this position that More argues for what is here being called "hard transhumanism," in that it is a transhumanism that tends to argue from a platform of a reductionist resolution *as well as* analysis. As has been shown, reductionism is the avowed foundation of Parfit's *analysis* or *method* of inquiry. And Parfit, by his own description, persists in describing as reductionist his own *resolution* of unproblematic nondeterminism—that is, his observation concerning empty questions. But to what extent do others agree with Parfit in terming such a resolution a reductionist one? The remainder of this chapter will be dedicated to placing Parfit's *empty question* in a much broader context. It seeks to provide evidence to suggest that More's philosophy represents a hard transhumanist outlook while

Parfit's is significantly different. And at the same time, there are process philosophies that share a common vocabulary with both of the former, while not arriving at a final *resolution* of reductionism, even while deploying it as a method. One of the oldest such process philosophies—and one with which Parfit himself has a very limited engagement that invites analysis and expansion—is the range of process philosophies pursued by various thinkers in the philosophical schools of Buddhism.

Mistaken Identities: Just How Buddhist Is Parfit?

In *Reasons and Persons*, Parfit cites the Buddha as a close ally, though this is a connection he later repudiates. Parfit's Buddha of *Reasons and Persons* arrives at two key moments. The first moment overcomes Parfit's own "disturbing" anxiety over the possibility that Parfit's method of reductionism is not something applied to "all people at all times" but is instead culturally delimited. The Buddha's purportedly sympathetic view is offered as strong evidence that Parfit's claims are not culturally bound, but are more objective observations that are not necessarily the result of Western lineages of thought.[40] Parfit's Buddha reveals Parfit's philosophy as objective and not merely culturally determined. The second moment for Parfit's Buddha is his and Parfit's shared claim, *pace* Nagel, that believing the reductionist view is not impossible, albeit very difficult. For Parfit, to countenance the reductionist method in its conclusion that personhood-over-time is an *empty question* involving no *further fact* or Cartesian self is certainly difficult but, when successfully done, results in the salutary triumph of the rational over doubts or fears, which are seen to be irrational. Such fears are lost, Parfit has claimed, "not because Teletransportation is *about as good* as ordinary survival. It is because ordinary survival is *about as bad as*, or little better than, Teletransportation. *Ordinary survival is about as bad as being destroyed and Replicated.*"[41] Parfit allows that he agrees with the Buddha that belief in a reductionist view is possible, given effort. Thus Parfit's Buddha here reveals the pragmatic viability of Parfit's philosophy. These two very limited engagements—a few pages within a 500+ page book—are the initial basis for Parfit's own self-avowed connection with Buddhism.

State of the Field: A Partial Review of the Buddhist-Parfitian Comparative Literature

What of Parfit's later repudiation of his affinity with the Buddha? After

Reasons and Persons, Parfit reverses course, attempting to disambiguate his and the Buddha's philosophy. Parfit terms his own outlook a Constitutive Reductionism with respect to persons, vs. the Buddha's Eliminativist Reductionism. But such an assertion is widely held to be incoherent, as will be shown below.

As is evident to perhaps any scholar of Buddhism, Parfit's engagement with questions of personhood and mereological analysis alongside ethical considerations grant him much greater common philosophical vocabulary with Buddha than the Buddha's brief mentions in *Reasons and Persons* might indicate. Accordingly, a substantial body of study has arisen analyzing the relationship between Parfit's analysis and Buddha's—since even before *Reasons and Persons,* but especially after the turn on the century. Steven Collins was likely the first to cite Parfit in connection with a discussion of Pāli Theravāda Buddhism, as early as 1982.[42] This, in spite of the fact that Parfit had not yet touched upon Buddhist ideas himself. Collins' work, together with Parfit's *Reasons and Persons*, drew the attention of Matthew Kapstein in a 1986 review. In his excellent 1990 dissertation, Nigel Tetley investigates the reincarnation issue in Theravāda Buddhism, seeking to demonstrate that a survival-identity dichotomy exists in Parfit that is not be found in early Buddhism, and that such Buddhism is thus "the product of a limited reductionist thesis (in the sense that it is a less radical thesis than Parfit's)."[43] Theravāda Buddhism, it is claimed, operates not along the lines of *Relation R*, but in terms of causal processes to do with action-and-consequence relationships (*karma*), which Tetley describes in terms of what he calls *Relation K*. In 1997, Collins also argues from *Relation K* in characterizing Theravāda Buddhism as less reductionist than Parfit.[44] Paul Williams' 1998 book extends Buddhist–Parfitian comparative work to later Mahāyāna Buddhism in a critical philosophical treatment of the Mādhyamika ethicist Śāntideva, which also makes some use of arguments from reincarnation.[45] Such treatment has been subjected to constructive criticism from Mark Siderits, who moves to defend Śāntideva against some of Williams' critiques. Such arguments concerning reincarnation are noted here by way of background, but do not concern the present chapter in the main, as doctrines of reincarnation generally fall well outside the consensus of Western scientific credibility.[46] Such arguments, then, have recourse to epistemic warrants (*pramāṇa*s) that make for—to say the least—awkward transdisciplinary comparison.

What of analyses for which *Relation K* arguments are not a key feature? Around the turn of the twenty-first century, Siderits has made efforts to lend some terminological clarity to discussions around Parfit, Buddhist doctrines, and eliminativist vs. reductionist characterizations thereof.[47] *Pace* Jim Stone and James Giles, to whom he directly responds,[48] Siderits sees the Buddhist position as not at all eliminativist (as Parfit claims) with respect to persons, but rather a "consistent and complete Reductionism."[49] Siderits here responds also to Parfit himself, who by this time had (in apparently partial agreement with Stone and Giles) seen the need to offer a fine distinction between his own philosophy—which he puts forward as a Constitutive Reductionism—and the Eliminative Reductionism he sees embodied in certain Buddhist texts. Parfit's own philosophy is constitutive, he tells us, in the sense that "On this view, though a person is distinct from that person's body, and from any series of thoughts and experiences, the person's existence just *consists* in them."[50] This is contrasted with an eliminativist perspective, with the Buddha's own view characterized as, "There really aren't such things as persons: there are only brains and bodies, and thoughts and other experiences."[51] For Siderits, by contrast, it is evident that the Buddha's perspective does not eliminate talk of persons in the way that, say, a secularized society with a germ theory of disease entirely dispels all talk of evil spirits as the cause of disease.[52] But, rather, talk of persons is retained in deference to convention and language usage. In the language of the Buddhist distinction between absolute and relative truth the Buddhist position is not, for Siderits, one that argues against the existence of persons on a *relative* (in addition to an absolute) level of truth. In brief, Parfit contrasts the constitutive with the eliminative as degrees of reductionism. For Siderits, "Reductionism is a middle path between the extremes of Non-Reductionism and Eliminativism."[53]

More recently, Jonardon Ganeri makes a critique of Parfit's characterization of the Buddha that is similar to that of Siderits, but Ganeri does so on philological grounds.[54] In sum, amongst Parfit's repertoire of Buddhist quotes there appears repeatedly a quote from the *Sutta on Ultimate Emptiness*, which does, as Parfit presents it, appear to take an eliminativist stance with respect to persons.[55] Ganeri demonstrates how this very quote, taken by Parfit from an English translation of a Tibetan rendering of the Sanskrit, has distorted the much more likely meaning of the original Sanskrit passage as it has informed Buddhist tradition. The Buddha, as portrayed here, is not saying that there is no such thing as the person, but rather that there is no person independent of its constituent parts, the famous five aggregates (*skandhas*), of which the person is composed. In other words, the original Sanskrit passage seems, on its own, to communicate not an eliminativist but a constitutive reductionist view (which would not differ substantially from Parfit). Ganeri cites the English translation of Duerlinger, from the *Sutta on Ultimate Emptiness* quote, taken from a famous commentary by the influential Buddhist epistemologist, Vasubandhu:[56]

> asti karma, asti vipākaḥ, kārakas tu nopalabhyate. ya imāṃś ca skandhān nikṣipati, anyāṃś ca skandhān pratisandhāty anyatra dharmasaṅketāt.[57]

> There is action and its result, but no agent is perceived that casts off one set of aggregates and takes up another elsewhere apart from the elements agreed upon (to rise interdependently).[58]

Moreover, Vasubandhu deploys this quote in an argument against the doctrine—widespread in his day—of the Pudgalavādins. Such Buddhists accorded a special status to the person (*pudgala*), and gained enough public traction with their arguments that Vasubandhu evidently felt the need to refute them. Ganeri compares these several viewpoints in the following way:

> Parfit distinguishes, among various sorts of ontological dependence, between adjectival and compositional ontological dependence. The first is illustrated by the relationship between a dent and its surface; the second by the relation between a tree and the cells of which it is composed. The difference between the Pudgalavādin and Vasubandhu might then lie in this: the Pudgalavādin thinks that persons are adjectivally dependent on streams of physical and psychological elements, while Vasubandhu thinks that the relation is one of compositional dependence (a person, Vasubandhu says, is not other than the stream).... If this way of distinguishing views is right, then both have given textually admissible interpretations of the Buddha's words, and neither endorses the Cartesian view; but only Vasubandhu is a Reductionist. None of our Buddhists, though, hold what Parfit has called 'the Buddhist View'. Indeed this is *ucchedavāda*, nihilism, one of the two extremes between which the Buddha sought a Middle Way, the other being *śāśvatvāda*, eternalism, the Cartesian View.[59]

That Parfit has misapprehended his philosophy's relationship to Buddhism should come as little surprise, involved as he is in a monumental project of his own, and being not at all a specialist in Buddhist philosophy. It is then, at present, far less interesting to say that Parfit imperfectly understands Buddhist thought—this being amply demonstrated. Rather, it seems clear that both Parfitian and Buddhist thought partake of the vocabulary of what Parfit calls a constitutive reductionism, and are involved in subtle arguments. That Parfit's view should accord with Buddhism ought to excite rather than discourage Parfit's followers—being as it is not evidence of Parfit's unoriginality so much as it is an instance of Modern and Premodern traditions sharing a common vocabulary. But what, precisely, is the extent of such accord, and how might they relate to one another? And how does transhumanist philosophy figure into this? Such questions will no doubt remain open, but one possible overview is offered below.

We have seen that Parfit's recent articulation of his type of reductionism entails a constitutive, as opposed to eliminativist, reductionism with respect to persons. Moreover, Parfit's formulation of what constitutes personhood over time specifies physical and psychological continuity and connectedness with the right kind of cause. This latter stipulation is, of course, intentionally open to interpretation. Parfit's system is, in this sense, intentionally flexible and inclusive. Such a situation has resulted in a hard transhumanist appropriation of Parfit's thought. Self-avowedly *functionalist* with respect to persons, hard transhumanism—as articulated by More—pushes Parfit's "right kind of cause" stipulation to an extreme by circumscribing it solely in terms of intentionality. That is, personhood persists insofar as there remains intentionality driving the physical and psychological continuity and connectedness that instantiates the person over time. Moreover, current transhumanist discussions around strong AI and personhood often concern situations in which there is little-to-no physical aspect. Thus, for hard transhumanists, Parfit's formulation is challenged not only by being anchored in eliminativist extremes, but the physical continuity and connectedness stipulation of Parfit is de-emphasized, if not ignored—as in the case of "substrate-free" persons. In sum, Parfit's philosophy alone is sufficient to account for a type of "classical" transhumanism. Selective normative developments upon Parfit have resulted in a sort of extreme Parfitianism—a phrase more or less synonymous here with hard transhumanism. On the present theory, Parfit's view represents what has become something of a classic transhumanist view, with the extreme formulation of More and others representing an influential hard transhumanist outlook. How does the literature of Buddhist philosophy—comparatively, much more vast—fit in?

SECTION 3: CONNECTING THE DOTS: BUDDHIST RESPONSES TO POSTMODERN PROBLEMS

As mentioned above, Parfit's reductionist, deterministic analysis of persons leads him to a resolution of an unproblematic nondeterminacy with respect to persons. When we have accounted for *Relation R*, we are in possession of all the facts. Whether the duplicate is or is not the same person remains an empty question. The interpretation of Parfit's *Relation R* hinges upon causality. His phrase "the right kind of cause" has been used by More and other Parfitian transhumanists to express the idea that direct intentional causation should be enough to consider a duplicate person equivalent to the original person. Wide-ranging though Parfit's theory is, causality remains one of its central concerns.

Of particular interest, then, in the dialogue between transhumanists and Tibetan Buddhists is the intellectual primacy the latter group places on the Indian philosopher Nāgārjuna, who was active in the first centuries of the common era. To do justice to a contrast between Nāgārjunian and Parfitian thought would be the stuff of a separate study. In brief, Nāgārjuna's philosophy is of immediate interest due to his famous skeptical fourfold negation, which is itself a deconstruction of causality. His "Root Verses on the Middle Path" (*Mūlamadhyamakakārikā*s) opens with the following:

> Neither from itself nor from another,
> Nor from both,
> Nor without a cause,
> Does anything whatever, anywhere arise.[60]

The interested reader is enthusiastically commended to Garfield's monograph from which the above translation—from the Tibetan—is taken. To paint in broad strokes—here Nāgārjuna advances what is intended as a delicate mediation between the extremes of essentialist and nihilist understandings of causality. That is, he seeks to deconstruct the notion of some intrinsic essence inhering in either a cause or its effect, while also critiquing any assessment of reality as nihilistically nonexistent. Nāgārjuna's famous *śūnyatā* (Skt.), often glossed "emptiness," is a subtle proposition, and entails his famous doctrine of "two truths." It is not that the relative level of apparent objects is devalued against the backdrop of a valorized ultimate level of emptiness. Nor is it the case that *śūnyatā* is itself to be understood as a hypostasized or reified condition. Rather, says Nāgārjuna, objects of sense perception are less the product of inherent causal essences and more the outcome of decentralized, evanescent conditions or flow states.

Garfield's own exegesis on the philosophical core of Nāgārjuna's Root Verses: its 24th chapter, may be of particular interest to the transhumanist discussion, and is worth quoting at length:

> This, of course, is the key to the soteriological character of the text: Reification is the root of grasping and craving and hence of all suffering. And it is perfectly natural, despite its incoherence. By understanding emptiness, Nāgārjuna intends one to break this habit and extirpate the root of suffering. But if in doing so one falls into the abyss of nihilism, nothing is achieved. For then action itself is impossible and senseless, and one's realization amounts to nothing. Or again, if one relinquishes the reification of phenomena but reifies emptiness, that issues in a new grasping and craving—the grasping of emptiness and the craving for Nirvāṇa—and a new round of suffering. Only with the simultaneous realization of the emptiness, but conventional reality, of phenomena and of the emptiness of emptiness, argues Nāgārjuna, can suffering be wholly uprooted.

Nāgārjuna's intellectual project presupposes the anti-essentialism endemic to the Buddhist philosophical heritage, and tries to account for such anti-essentialism without succumbing to nihilism. Nāgārjuna is, nonetheless, often accused by essentialists as espousing a type of nihilism. Or as we have seen, in Parfitian terms, is misunderstood (along with Buddhism in general) as advancing a reductionist-eliminativist doctrine with respect to persons. However, Nāgārjuna seeks to run a middle course between essentialism and nihilism—or in Parfitian terms, a middle course between the "further fact" mentality of the Cartesian ego, and a reductionist-eliminativist doctrine. For Nāgārjuna, persons are 1.) empty of inherent existence and possess only relative existence, and are 2.) not the product of inherent causality, but of aggregated conditions. It is not the argument of this chapter that a Nāgārjunian outlook is without potential difficulties of its own. After all, the reifying tendency of human beings remains rather stubborn. Rather, the Nāgārjunian outlook may mitigate against one of the great tragedies of hard transhumanism: the reification of a functionalist outlook on selfhood, at the expense of a type of nihilism with respect to the embodied.

The Bizarre Love Triangle: Parfit, More, and Buddhism under a Transdisciplinary Lens

Two more complementary perspectives are worthy of special note in addition to the discussion thus far. Roy Perrett, to begin with, has applied to the Buddhist-Parfit discussion the Minimalist arguments of Mark Johnston.[61] Johnston articulates a Minimalism with respect to personhood that can be summarized thus: An account of the metaphysical deep facts of personhood by themselves generally accomplish very little in terms of normative behavior.[62] Even if true, a constitutively reductionist account of persons is often minimalized in its social importance—or, at best, works against the inertia of a "commonsense" nonreductionist account of persons as bounded, separate, independent entities. *People* in general, then, are unlikely to be fast in shifting to the views that Parfit prescribes, even if such views were more "rational" or "correct" on some level. Perrett's deployment of Johnston suggests that there exists a social perception of persons as discrete objects that largely persists in spite of the deeper philosophical reality of persons as processes.

In a parallel insight from the discipline of neurology, Michael Graziano has put forth his attention schema theory, in which "awareness is a schematized, descriptive model of attention." Graziano's theory, if correct, has some explanatory power concerning social adherence to illusory phenomena. For example, 'purity' associations around the color white are undiminished knowing that the color is, in reality, a muddy mixture of all colors, because "white is not represented in the brain as a mixture of colors but as luminescence that lacks all color."[63] Perceptual truths continue to have traction, even when deeper facts are well-known. Graziano also takes care to distinguish between awareness attributed to oneself (Awareness A) and awareness attributed to another (Awareness B).[64]

Such a distinction might also be of great usefulness in the personhood discussion, as it is a distinction that Parfit's theories do not account for.

Ethics and Morals: The Long Hard Road out of Suffering

Broadly speaking, Buddhist philosophy partakes of a vocabulary akin to that of Parfit's. *Pace* Parfit's own characterizations of Buddhism, its formulations are generally much more aligned to some form or other of constitutive reductionism, by virtue of consciously avoiding the extremes of eternalism (venerating something like a Cartesian ego) and nihilism (denying the persistence of persons entirely). What precise forms that constitutive reductionism might (or might not) take could easily be the subject of another study. Within the intentionally narrow consideration of the question of personhood, it might seem that both Buddhist and Parfitian discourse represent strongly overlapping fields of philosophical consideration, with Morean hard transhumanism at the extreme end. Should the focus of inquiry widen somewhat to include ethics, however, the picture rapidly gains complexity. The present author hopes to treat this topic at a later time. For the present, however, a few notes on the subject would not be out of place.

As articulated by the Buddha's famous Four Noble Truths, Buddhism is foundationally concerned with a liberatory path out of life's inherently dissatisfactory conditions. Since the advent of the Mahāyāna, the locus of concern, as articulated in the bodhisattva vow, extends to all sentient beings (including, crucially, the bodhisattva herself). Such a view presupposes persons, while also being committed to the experiential ascertainment of the person's nonessentiality as an important facet of the path of liberation. Parfit, for his part, is motivated by subjecting Lockean and Consequentialist ethical questions to some rigorous interrogation. In doing so, he seeks to develop an account of persons that in no way presupposes their essentiality. Concluding that persons just consist in *Relation R*, he finds himself very much the better for the certainty of his knowledge, being, as a consequence, less exclusively self-concerned. More, for his part, has in mind an emancipatory agenda that is notably more individualist in scope, concerned as it is with human betterment through individually willed evolutionary exploration. The Parfitian-Morean connection is clear from the beginning. Much of this chapter has been devoted to exploring the Parfitian-Buddhist connection. But it is when the hard transhumanist ethics of personal evolution and morphological freedom are contrasted with the bodhisattva orientation that the rubber really hits the road—or in Buddhist terms, the path. For transhumanists and Buddhists have begun to cooperate with one another—each having salutary aims, but from opposite ends of the park—individualist and collectivist. There is, then, a tension-of-opposites that lends potential creative fire to this partnership. Transhumanists, in partnership with Buddhists, may well explore the achievement of their aims fruitfully. But, from a Buddhist perspective, what is achieved may not be at all what was expected.

Human Evolution: The Next 100 Years

Without a doubt, the coming century will see significant changes to the conventional understanding of personhood, and what it means to be human. Central to the human evolutionary process, then, will be a flexible and robust understanding of ethics that keeps pace with the rapidity of technological and social change. A look at ethics in Parfitian proto-transhumanism and Buddhism certainly warrants further study. It is to be hoped, it must now suffice to say, that the next step in humanity's evolutionary process is as much an ethical step as anything else. Taking the foregoing study into account, how might such an ethics begin to take shape?

Buddhist discourse, it must be remembered, takes as its core assumption the reality of basic dissatisfaction and the specific means to its mitigation. Increasingly with Mahāyāna Buddhism, there is an emphasis upon mitigating the basic dissatisfaction of all beings—oneself and others—with great impartiality. In this regard, the Mahāyāna is a strong forebear and exemplar of an ethical impulse guiding modernity: that of the widening of ethical concern beyond one's default culturally determined horizon. In contemporary discourse, this includes an outgrowth of compassion beyond national, regional, and political boundaries. It includes an honoring of the many different manifestations of social diversity—ethnic, cultural, class, gender, sexual, cognitive, and bodily diversity. It includes a sincere extension of concern beyond one's own species. And, crucially, it includes an impartiality with respect to the deployment of compassion. That is, the perpetrator as well as the persecuted both fall within the ambit of compassionate action and intention.

Such noble intentions will no doubt be tested to the extent that transhumanist aims are realized. Even without fervent transhumanist support, certain technological developments seem almost inevitable, any one of which might put the above program of impartial compassion, the bodhisattva orientation, to the test. For example, it may become increasingly less controversial to locate the Buddhist category "sentient being" in an artificial organism or cybernetic human being. But is a sentient being to be located in the substrate-independent consciousness of transhumanist literature? That is, in a strong AI unconnected to a particular body? If so, would such an AI person be in possession of Buddha-nature? A potential teacher of Dharma? Or even a reincarnated master (*tulku*)?

These questions have no easy answers, and ought to be discussed by qualified Buddhist teachers of the future, not by academics. But posing such questions may help illustrate a type of counter-reaction against transhumanism held by many human beings. Hard transhumanists may envision their program as one that uplifts humanity's future, save for a few "neo-primitivist" luddites here and there. But for many, the success of the transhumanist program might very well lead to a specious "evolutionary" step, in which transhumanists are alienated from and subjugate nontranshumanists in a way analogous to the widespread subjugation of nonhuman animals by human animals, the occasional animal rights advocate notwithstanding. In the view of others, the transhumanist project may be inherently self-destructive, so narrowly focused on the global elite availing itself of transhumanist achievements that the vast majority of human beings will never make the leap from human to transhuman status.

Whatever future we speculate, the contention of this study is that a certain transcendentalist teleology has been guiding the formation of transhumanist thought. The hard transhumanist project appears to ask: What technological project or achievement would help displace an "imaginary" religious transcendentalism and substitute in its place a "real" technologically based transcendentalism? It appears to ask: What technological means would allow humanity to escape its sad, mortal, physically embodied condition?

To this question, transhumanism supplies agendas such as the Global Future 2045 "Avatar Project" agenda, in which the imminent (within the next decade) transplantation of the brain to an artificial body will be surpassed by artificial persons with artificial brains and finally a "hologram-like avatar." In contrast, something like a Nāgārjunian Buddhism would address the above question (concerning technological means) by simply saying: "None!" That is, no technology is capable of obviating the four noble truths. Certain elements of the human condition—birth, sickness, old age, death—are calamities that invite mitigation, catalysts that impel salutary effort, and immutable defining facts in the existence of sentient beings. Even technologies that slow or reverse senescence do not remove the pain of birth and the thousand natural shocks of embodied existence. Even if one desired it, prolonging one's likeness in another body is less a key to immortality than a way to simply, tautologically... propagate one's likeness! This is a trick that evolutionary biology has long since supplied to some degree in the form of offspring. And it is a trick that has far from served to undo the urgency and applicability of the four noble truths.

But specifically in Nāgārjuna's view, I would argue that there is to be found neither a prohibition against, nor a sanction of, the idea of *Relation R* or transhumanist deployment of this idea. The artificial person that succeeds me, in this view, is in an unproblematic state of nondeterminacy.

We cannot definitively say that such a person is me, is not me, is both, or is neither. We can say only that such a being is the product of conditions, not one in which my identity inheres. How this translates into any sort of legal status remains to be seen. Does my "closest continuer" retain all of my legal rights? However this question is to be answered following a Nāgārjunian philosophical reading, the important thing is that there is greater depth and nuance than the hard transhumanist response: "Well, yes, of course!"

EIGHT

Shared Vulnerabilities:

Cosmic Consciousness and the Philosophy of Organism

Jason James Kelly

"The whole of Creation is *alive*."
-Edward Carpenter

IN HIS CLASSIC ESSAY, "The Historical Roots of the Ecological Crisis" (1967), the American historian Lynn White Jr. claims that the historical roots of the ecological crisis can be traced to the dominant role Western religion played in shaping the direction of the modern scientific method. In particular, White suggests that this religious worldview privileges an anthropocentric attitude that justifies "man's" dominion over nature. In turn, this anthropocentric attitude shaped the ideological contours that eventually gave rise to the scientific method and the technological exploitation of nature. If White is correct, then it is a mistake to assume that science can save us from the ecological crisis because its fundamental axioms are rooted in an anthropocentric interpretation of nature. White's position on the matter is clear: "more science and more technology are not going to get us out of the present ecological crisis."[1] The contemporary environmental author Kenneth Brower shares White's concerns when he writes, "the notion that science will save us is the chimera that allows the present generation to consume all the resources it wants, as if no generations will follow. It is the sedative that allows civilization to march so steadfastly toward environmental catastrophe. It forestalls the real solution, which will be in the hard, nontechnical work of changing human behavior."[2] The former president of the Yale School of Forestry and Environmental Science, James Gustave Speth, seems to identify a similar problem when he notes, "our environmental discourse has thus far been dominated by lawyers, scientists, and economists. Now, we need to hear a lot more from the poets, preachers, philosophers, and psychologists."[3] In other words, we need to make more of an effort to expand our understanding of the human-nature relationship to include a deeper engagement with the humanities. To this end, perhaps new religious narratives need to be written that recognize the *intrinsic* value of nature, for only then will we be inclined to alter our destructive behavior and create a new course of social action based on the environmental principles of stewardship and sustainability.

In the past few decades this challenge has been taken up by a growing field of research called spiritual ecology. Generally speaking, spiritual ecology focuses on the study of spiritual narratives that highlight nature's intrinsic value and aim to unify the ontological divide between the material and spiritual domains.[4] The philosophy underlining spiritual ecology is based largely on the philosophy of deep ecology, which rejects anthropocentric interpretations of nature for a more ecocentric viewpoint and that privileges such notions as holism, relationality, and interdependence.[5] A central assumption shaping both deep ecology and spiritual ecology is that "self-realization" is ultimately a product of our intuitive identification with nature. The essential difference between these two schools of thought is that spiritual ecology makes explicit what is implicit in deep ecology—namely, that this intuitive identification with nature has spiritual significance. As the name indicates, spiritual ecology places spirituality at the forefront of its philosophy and focuses on how spiritual narratives have been and continue to be utilized by a wide range of spiritual traditions—from traditional Indigenous knowledge and Eastern philosophies to ancient paganism and new-age spiritualties—to enhance our relationship with nature. What all these spiritual traditions hold in common is the experientially based belief that our connection with nature is spiritually valuable. While the field of spiritual ecology has contributed greatly to enhancing our understanding of the diverse ways in which spiritual traditions address ecological issues, I argue that it has largely overlooked a particular lineage of ecological thinking in the West that emerged in the late nineteenth and early twentieth centuries and that focused on developing a new way of relating to nature called "cosmic consciousness." I suggest that the concept of cosmic consciousness, which I characterize as the experiential awareness of one's connection to or unity with the cosmos, can be considered a valuable alternative for thinking about how humans relate to nature.

Given the material fact that "we are born of stars and made of Earth"[6] it perhaps only makes sense that we seek to spiritualize our connection with the cosmos. It could perhaps even be argued that the desire to realize cosmic consciousness and become "One" with the forces of nature is at the root of all religious expression. Yet the move to characterize cosmic consciousness as an ecological concept has a distinct, largely unacknowledged history that can be traced to the teachings of two intellectuals of the late nineteenth and early twentieth centuries: the English author and activist Edward Carpenter and the Canadian psychiatrist Richard Maurice Bucke. Carpenter's and Bucke's understanding of cosmic consciousness was inspired in large part by the teachings of their shared mentor, the American poet Walt Whitman, who envisioned the rise of a radical new way of *being-in-the world* that embraced the findings of both science *and* spirituality. The move to construct this link suggests a connection between the history of cosmic consciousness and the philosophy of Alfred North Whitehead. This connection raises some interesting questions. How might the spiritual and scientific conception of cosmic consciousness compliment the metaphysical system underlining Whitehead's philosophy of organism? How might the overlap between cosmic consciousness and the philosophy of organism enhance our contemporary understanding of the ecological crisis? Is it possible that a deeper study of the relationship between cosmic consciousness and the philosophy of organism could help to expand our understanding of the human-nature relationship?

In this essay I argue that the concept of cosmic consciousness and the philosophy of organism privilege a nondualistic or "holistic" conception of the subject-object relationship that affirms an ecological relationship with nature. I begin with the history of cosmic consciousness in the work of Carpenter and Bucke, with the aim of demonstrating its spiritual and scientific significance. Next, I draw on the teachings of spiritual ecology to demonstrate how cosmic consciousness can be understood as an ecological principle. Finally, I show how the ethical significance of cosmic consciousness can be enhanced by cultivating a deeper engagement with Whitehead's philosophy. Stated simply, I suggest that Whitehead's philosophy provides a metaphysical foundation to ground the ethical teachings of cosmic consciousness. By elucidating the overlap between these two, I hope to demonstrate how such a cross-fertilization of ideas can help forge a new path of ecological sustainability for future generations.

THE SPIRITUAL SIGNIFICANCE OF COSMIC CONSCIOUSNESS

The term "cosmic consciousness" was first coined by the English author and activist Edward Carpenter (1844–1929), in his work *From Adam's Peak to Elephanta* (1895), which details his experiences traveling through Ceylon (Sri Lanka) in 1890 with the hope of learning about the mystical teachings of Hindu philosophy. Carpenter eventually cultivated a relationship with a "Gnani," or "wise man," named Ramaswamy who taught him some of the fundamental beliefs of Hindu mysticism. According to Carpenter, "what the Gnani seeks and obtains is a new order of consciousness—to which, for want of a better, we may give the name *universal* or *cosmic* consciousness, in contradistinction to the individual or special bodily consciousness with which we are all familiar."[7] And it is here, in the context of an English foreigner describing the mystical heights of Hindu philosophy, where we find the first direct reference to this concept. Carpenter goes on to describe cosmic consciousness as an illuminative state of "non-differentiation" that blurs the boundary between "I" and "Other."[8] The "chief fact" of cosmic consciousness, according to Carpenter, "is not that you are distinct from others, but that you are a part of and integral with them."[9] Drawing on his own interpretation of Hindu philosophy, Carpenter claims that this cosmic bond between "I" and "Other" leads to the moral conviction that "the divine spirit" resides in every creature and in nature as a whole.

Edward Carpenter was truly a man ahead of his time. He was a staunch socialist who fought tirelessly to call the public's attention to various progressive causes, including the labor movement, homosexual rights, gender equality, prison reform, animal rights, vegetarianism, and even early forms of environmentalism.[10] His interest in supporting these controversial issues is all the more remarkable given that he was born into privilege and benefited from all the economic advantages that came with being part of an upper-middle-class household in the Victorian era. He graduated from Cambridge in 1868 and gave up a prestigious position as a curate to participate in the university extension program, which sent teachers to rural areas of England to teach the working class. It was while living amidst the poor farmers of Northern England, far from the comforts of Cambridge, that Carpenter's eyes were first opened to the unjust conditions caused by economic and social inequality. At the same time, Carpenter was also inspired by the simplicity of country living and by the romantic idea that working intimately with the land could instill a sense of harmony with nature—a harmony that Carpenter believed was slowly being eroded by the dominance of "commercial civilization." After inheriting a substantial sum of money, Carpenter was free to retire from teaching and concentrate his efforts on his greatest passion: writing poetry. He bought a small farm near Sheffield and began composing his most famous book of poetry, *Towards Democracy* (1883).

Carpenter describes *Towards Democracy* as the "start-point and kernel of all my later works, the center from which the other books have radiated."[11] In many ways this is an accurate assessment, because the book signals Carpenter's first attempt to express his views on mystical experience, which became the core teaching of his entire philosophy.

In 1881, Carpenter's life was transformed by three events: his mother died, he obtained a copy of the Bhagavad Gita, and he had his first mystical experience. Carpenter describes his encounter with the mystical as follows: "all at once I found myself in touch with a mood of exaltation and inspiration—a kind of super-consciousness—which passed all that I had experienced before, and which immediately harmonized all these other feelings, giving to them their place, their meaning and their outlet in expression."[12] The loss of his mother, combined with his exposure to Eastern philosophy and his own realization of "super-consciousness" helped crystalize Carpenter's desire to write *Towards Democracy*. But there is still one other source of inspiration that cannot be overlooked, namely, Walt Whitman's *Leaves of Grass* (1855).

Walt Whitman (1819–1892) was an American poet who dreamed of creating a new spiritual vision for American society based on the democratic ideals of equality and comradeship. Similar to his mentor, Ralph Waldo Emerson, Whitman believed that our capacity to spiritually identify with the "Kosmos" was key to unlocking our moral potential, and once this moral potential was collectively realized, society would be transformed for the better. However, unlike Emerson and the other Transcendentalists, Whitman placed much more value on the erotic dimensions of this mystical bond with the "Kosmos." In fact, Whitman was convinced that "there was a close connection—a very close connection—between the state we call religious ecstasy and the desire to copulate."[13] Whitman did not believe that erotic desire was an obstacle to spiritual growth; on the contrary, the pleasures of the body and erotic desire in general can be utilized as a catalyst to unleash the full power of our spirituality. In other words, Whitman was convinced that erotic desire and spiritual desire were two sides of the coin and thus he famously declares himself, "poet of the Body and poet of the Soul."[14] The core spiritual teaching of *Leaves* is the idea that erotic desire can act as mystical bridge linking the profane (body/nature) to the sacred (soul/God), and vice versa.

Carpenter was fascinated with Whitman's mystical vision and particularly drawn to the erotic subtext of *Leaves*. This is unsurprising, given that both Carpenter and Whitman were homosexuals who struggled to evade social condemnation and find some sort of fulfillment with their erotic desires.[15] It's quite likely that Carpenter sought out Whitman as a mentor in order to make sense of this link that Whitman identified between the spiritual and sexual domains. Carpenter eventually struck up a friendly correspondence with Whitman and visited him once in 1877 and again in 1884. Overall, Whitman was impressed with Carpenter and encouraged his writing. For his part, Carpenter viewed Whitman as a kind of prophet preaching a new "Kosmic" gospel of universal love. In his book, *Days with Walt Whitman* (1906), Carpenter characterizes Whitman as follows:

> In many ways Whitman marks a stage of human evolution not yet reached, and hardly suspected, by humanity at large; but in no respect is this more true than in respect of his capacity of Love. If you consider Whitman's life you will see that Love ruled it, that he gave his life for Love. There were other motives no doubt, but this one ultimately dominated them all. It permeates like a flame his entire writings; it took him to the battlefield and the hospitals in succor of the wounded soldier; it led him (before the war deep into the life and comradeship of the people)—all phases; after the war it united him in bonds of tender and life-long friendship with many, both men and women; it surrounded his death bed with devotion, and brought thousands to his funeral. For it he gave away his possessions and material means of life, he gave his prospects of professional success, he gave health, fame—all that a man can give—and accepted illness and obscurity, and oftentimes long and painful loneliness and betrayal even of love itself.[16]

It was love, then, a love of both body and soul, a love of both God and nature, that Carpenter believed was key to understanding Whitman's mystical vision. And it was precisely this type of nondualistic mystical vision that Carpenter sought to carry on in his own Whitmanesque book of poetry, *Towards Democracy*.

Similar to Whitman's famous poem, "Song of Myself," Carpenter's *Towards Democracy* can be read as a kind of mystical treatise describing the soul's erotic ascent and eventual unification with God. But the journey begins with nature: "The sun, the moon and the stars, the grass, the water that flows round the Earth, and the light air of heaven: To You greeting. I too stand behind these and send you word across them."[17] Carpenter urges his reader to identity with the "magnificence and splendor of nature" and send "passionate kisses back to the sun!" For like Whitman before him, Carpenter believed that the material and spiritual domains are coextensive dimensions of one singular reality: "O the lifting of arms to Nature—heaven wrapped around one's body!"[18] By identifying with nature, in all its diversity and change, the soul learns how to grow, to expand, and this spiritual expansion breeds a new sense of moral responsibility that Carpenter describes in terms of "Freedom" and "Equality." Take, for example, the following passage:

> Do you understand? To realize Freedom or Equality (for it comes to the same thing)—for this hitherto, for you, the universe has rolled; for this, your life, possibly yet many lives—for this, death, many deaths—for this, desires, fears, complications, bewilderments, suffering, hope, regret—all falling away at the last duly before the Soul, before You (O laughter!) arising the full grown Lover—possessor of the password.[19]

Indeed, love is the password which unlocks the ultimate spiritual treasures. And, like Whitman, Carpenter's understanding of mystical love is thoroughly erotic: "Sex still goes first, and hands eyes mouth brain follow; from the midst of belly and thighs radiate the knowledge of self, religion, and immortality."[20] He goes on to proclaim: "The body is a root of the soul. As the body in air, so the soul sustains itself in love."[21] This nondualistic embrace of both body and soul engenders the ultimate form of mystical liberation: "O freed soul! Soul that has completed its relation to the body! O soaring, happy beyond words, into other realms passing, salutations to you, freed, redeemed soul!"[22]

In his next two major works, *Civilization: Its Cause and Cure* (1891) and *The Art of Creation* (1904), Carpenter begins to unpack the larger significance of cosmic consciousness in relation to both nature and social ethics. Drawing on a heavy mixture of Romanticism and German idealism, Carpenter presents a comprehensive view of the human condition that traces our ills to the accumulation of private property, which in turn precipitated our alienation from nature and the "common life." This failure on our part to sustain a spiritual connection to nature has impoverished our moral sensibilities, corrupted our views of the body, and blinded us to both social and ecological injustice. Thus, Carpenter calls for a return to nature, when "man will once more *feel* his unity with his fellows; he will feel his unity with the animals, with the mountains and the streams, with the Earth itself and the slow lapse of the constellations, not as an abstract dogma of science or theology, but as a living and ever-present fact."[23] But of course the question remains as to how we can achieve this return to nature? Carpenter's answer, unsurprisingly, rests on our capacity to realize cosmic consciousness.

Carpenter believed that consciousness has evolved through three stages of development. The first stage is "simple consciousness, in which the knower, the knowledge, and the thing known are still undifferentiated."[24] He refers to the second stage as "self-consciousness," which refers to a state of distinction between subject (knower) and object (known). The third and final stage is, of course, cosmic consciousness, in which "the object suddenly is seen, is *felt*, to be one with the self."[25] It is here, in this cosmic state of mystical illumination, where the artificial divide between mind and matter, body and spirit, nature and God, finally dissolves and a new sense of unity with creation rises to the take its place. Carpenter writes:

> This form of Consciousness is the only true knowledge—it is the only true existence. And it is a matter of experience; it has been testified to in all parts of the world and in all ages of history. There is a consciousness in which the subject and the object are felt, are known, to be united and one—in which the Self is felt to be the object perceived, or at least in which the subject and the object are felt to be parts of the same being, of the same including Self of all.[26]

Carpenter believed that this move from simple, to self, and eventually to cosmic consciousness was a product of Lamarckian evolution. Borrowing from Whitman, he refers to this process as "exfoliation," which he describes as "a force at work throughout creation, ever urging each type onward into new and newer forms. This force appears first in consciousness in the form of desire."[27] Carpenter was convinced that creation was alive and directed by an unseen desire that compels the energies of the cosmos to unfold in a particular way and for a particular purpose—namely, to help us realize both the breadth and depth of our love. And therein lies the essential ethical teaching of cosmic consciousness: we are obligated to care for each other, for nature, and for the whole creation, because from the perspective of cosmic consciousness, "thou art that" (*tat tvam asi*).

THE SCIENTIFIC SIGNIFICANCE OF COSMIC CONSCIOUSNESS

At the dawn of the twentieth century a new way of thinking about mystical experience began to emerge in the writings of certain intellectuals in the West that challenged the authority of organized religion by privileging naturalistic explanations of the mystical over theological ones. I am thinking in particular of two influential figures: the Canadian psychiatrist Richard Maurice Bucke and the American philosopher William James. The publication of Bucke's *Cosmic Consciousness: A Study in the Evolution of the Human Mind* (1901) and James's *The Varieties of Religious Experience* (1902) signal a watershed moment in the history of Western mysticism in that these two books radically changed the way mysticism was understood by both scholars and the public in general. Before these two works, the study of mysticism was largely dominated by theologians whose interest in studying mystical experience centered on reinforcing or validating the dogmatic claims of organized religion. In a revolutionary move, Bucke and James changed all this by exchanging theology for science. Thus their work laid the foundation for a new field of study: the psychology of mysticism.

In the *Varieties,* James suggests that mystical experience can be understood an "altered state of consciousness." Normal, "everyday," rational consciousness is but one mode of awareness. Under the right conditions, certain states of consciousness can arise that appear to be "discontinuous" with ordinary consciousness. James claims that these altered states of consciousness emerge from a "subliminal" or "subconscious" region of the mind that is usually foreclosed to rational consciousness. James believed that accessing these subconscious materials can change a subject's self-identity and, consequently, his or her worldview. In other words, accessing the deepest regions of the subconscious could produce a mystical state of consciousness. James identifies four defining characteristics of mystical consciousness: transciency, passivity, ineffability, and a noetic quality. He argues that these four characteristics "are sufficient to mark out a group of states of consciousness peculiar enough to deserve a special name and to call for careful study. Let it then be called the mystical group."[28] Of the four characteristics that James describes, the noetic quality plays a particularly important role in his overall model. The noetic quality refers to an intuitive sense of knowledge or "illumination" engendered by mystical consciousness. In psychological terms, "illumination" refers to a "highly specialized type of perception" characterized by an intense sense of "enlargement, union, and emancipation."[29] Ultimately, James reached the conclusion that mystical states of consciousness can produce an extraordinary kind of intuitive knowledge and that this knowledge is profoundly liberating.

Similar to James, Richard Maurice Bucke (1837–1902) was convinced that mystical consciousness can be understood in psychological terms. However, Bucke takes this scientific explanation a step further by suggesting that mystical consciousness is a product of evolutionary processes. According to Bucke, mystical consciousness is not only a natural phenomenon, but it also signifies a major advancement in the evolution of the human mind. This is the main argument shaping Bucke's most influential work, *Cosmic Consciousness* (1901). To this day, James's *Varieties* still overshadows Buck's *Cosmic Consciousness*, but it would be a mistake to underestimate the foundational role Bucke's study played in advancing our knowledge about the relationship between science and spirituality.

In 1838, when he was still just a child, Bucke's family immigrated to Canada from England. He grew up on a farm near London, Ontario, and left home at sixteen to find his fortune in the American Midwest. He eventually made his way to California, where he hoped to make a living as a prospector. However, his hopes were dashed in the winter of 1857, when he and his two companions became trapped by storms while trying to cross the Sierra Nevada mountain range. His two companions died and Bucke lost one foot to frost-bite. Shortly thereafter, he returned to Canada and eventually earned a degree in medicine from McGill University in 1862. A few years later he opened his own successful medical practice, and in 1877 he secured the position of superintendent for the London Asylum, which was one of the largest asylums in North America at the time. Bucke held the position for the rest of his life, which was tragically cut short in 1902, when, a year after he published *Cosmic Consciousness*, he slipped one evening while staring at the stars and fatally banged his head.

Bucke's personal interest in studying mysticism can be traced to an event that occurred on a spring night in 1872. After reading the poetry of Whitman and other Romantic poets late into the evening with friends, Bucke was proceeding home when suddenly he felt immersed in a flame-colored cloud. He describes the experience in the third-person as follows:

For an instant he thought of fire, some sudden conflagration in the great city; the next, he knew that the light was within himself. Directly afterwards came upon him a sense of exultation, of immense joyousness accompanied or immediately followed by an intellectual illumination quite impossible to describe.... Among other things he did not come to believe, he saw and knew that the Cosmos is not dead matter but a living Presence, that the soul of man is immortal, that the universe is so built and ordered that without any peradventure all things work together for the good of each and all, that the foundation principle of the world is what we call love and that the happiness of every one is in the long run absolutely certain. He claims that he learned more within the few seconds during which the illumination lasted than in previous months or even years of study, and that he learned much that no study could have ever taught. The illumination itself continued not more than a few moments, but its affect proved ineffaceable; it was impossible for him to forget what he at the time saw and knew, neither did he, or could he, ever doubt the truth of what was then presented in his mind.[30]

The profundity of this experience transformed Bucke, and he began dedicating his free time to studying the history of Western mysticism. What he discovered is that his experience is not that uncommon and that many individuals throughout history and across the world have described similar types of experiences. In fact, Bucke concluded from his research that mystical experience is at the root of all religion. However, Bucke was a critical thinker and found religious explanations of the experience quite unsatisfactory. He thus turned to science for an answer. In particular, he utilized his knowledge of science to construct a whole new way of thinking about the meaning of the mystical based on a unique combination of psychology and evolutionary theory.

Bucke borrowed the term "cosmic consciousness" from Carpenter and even adopted his developmental model, which consists of three stages of consciousness: simple, self, and cosmic.[31] Bucke and Carpenter were both inspired by the teachings of Whitman. In fact, their friendship began in 1880 when Bucke contacted Carpenter to obtain advice on a biography he was writing on Whitman. They exchanged correspondence for years after and even visited each other. Although they each shared a deep admiration for Whitman, their interpretation of his teachings varied considerably. For example, following Whitman, Carpenter made every effort to highlight the link between the spiritual and the sexual domains in his writings. Bucke, in contrast, never really addressed sexuality in any detail, even going so far as to strategically downplay the erotic themes of Whitman's poetry. Their interpretation of cosmic consciousness was also quite different. Whereas Carpenter focused his energies on understanding the spiritual and social relevance of cosmic consciousness, Bucke was primarily interested in elucidating its scientific significance.

Bucke begins *Cosmic Consciousness* by outlining his main theory that mystical or "cosmic" states of consciousness signify the evolutionary heights of the human mind, and that the attainment of mystical knowledge (illumination) places the individual on "a new plane of existence."[32] Similar to James, Bucke believed that illumination is the defining characteristic of cosmic consciousness, which is marked by an intuitive sense of "unity" with the universe. This intuitive knowledge, Bucke argues, produces a state of "moral elevation," a love for all existence that "shows the cosmos to consist not of dead matter governed by unconscious, rigid, and unintending law. On the contrary, it shows the cosmos as entirely immaterial, entirely spiritual and entirely alive; it shows that death is an absurdity, that everyone and everything has eternal life; it shows that the universe is God and that God is the universe."[33] Bucke claims that this "cosmic sense" is at the root of all spiritual expressions and the foundation of all religion. Socrates, the Buddha, Jesus, Muhammad, are all figures who experienced cosmic consciousness and the wisdom of their spiritual teachings are a product of this experience. But of all the religious prophets of the world, Bucke believed not one embodied the full potential of cosmic consciousness better than Walt Whitman. Bucke claims that Whitman's moral sensibility, his radical love and overall mastery of the spiritual domain, demonstrates "that he is the greatest case of cosmic consciousness to date."[34]

Bucke believed that experiences of cosmic consciousness were increasing and that someday in the future all humanity would evolve the capacity to enjoy its fruits. When that day comes, argues Bucke, the world will be transformed by the emergence of a new "cosmic" ethics based on the idea that "the foundational principle of the world is what we call love."[35] A lofty sentiment indeed, but Bucke's vision is undermined by his own racist and misogynistic biases. Bucke claimed, for instance, that cosmic consciousness "appears in individuals mostly of the male sex" and that the "Aryan race" possesses a superior intellect.[36] As tempting at it might be to try to rationalize Bucke's odious ideas as a product of the times, such a viewpoint minimizes the progressive nature of Carpenter's position, which claims that cosmic consciousness can be utilized to advance the cause of social justice. Nonetheless, despite the obvious limitations of Bucke's model, his overall aim to frame cosmic consciousness in evolutionary terms was a pioneering move that did helped bring spirituality and science into deeper discourse.

This dialogue that Bucke sought to construct between spirituality and science can perhaps be best understood in relation to the idea of "evolutionary mysticism." Both Bucke and Carpenter believed that there is a spiritual purpose to the evolution of life, and this idea is part of a long history of thought that is commonly referred to today as "evolutionary mysticism." The contemporary historian of religion, Jeffrey J. Kripal, defines evolutionary mysticism as "a philosophical system in which God's transcendence and immanence are simultaneously affirmed (*pan-en-theism*, literally, 'all-in-god-ism') and both find expression through the very process of cosmic evolution."[37] This notion of evolutionary mysticism shares a rich history with Western philosophy, as evident in the writings of such influential figures as Plotinus, J.G. Fichte, F.W.J. Schelling, G.W.F, Hegel, and Pierre Teilhard de Chardin, to name but a few.[38] At the beginning of the twentieth century and culminating with the counterculturral movement of the 1960s, the idea of evolutionary mysticism became more enriched by the teachings of various Eastern philosophies, transpersonal psychologies, and New Age spiritualties. At the heart of the contemporary understanding of evolutionary mysticism is the assumption that both dimensions of human consciousness—"the rational" and "the mystical"—are products of our shared evolutionary history, and that any comprehensive understanding of the human condition requires us to develop interpretive models that can accommodate the finding of both fields of knowledge. And it is precisely Bucke and Carpenter's identification with this type of "nondualistic" evolutionary framework that defines their shared understanding of the overlap between spirituality and science.

COSMIC CONSCIOUSNESS AND SPIRITUAL ECOLOGY

The concept of cosmic consciousness coheres well with the overall philosophy of spiritual ecology. The most obvious overlap between Carpenter and Bucke's teachings on cosmic consciousness and the philosophy of spiritual ecology is that both systems reject a dualistic conception of the relationship between "self" and "world"; that is, both systems privilege a nondualistic or "holistic" worldview. In addition, both systems also advance a conception of "self" that is expansive. This is important because it is precisely the self's capacity to expand that permits one to identify with nature/cosmos. Recall, that this capacity to identify "with the universe and everything in the universe"[39] is an essential characteristic of cosmic consciousness. Furthermore, both the teachings of cosmic consciousness and spiritual ecology suggest that the expansive self's nondualistic identification with nature/cosmos is revelatory in the sense that nature is no longer perceived as "dead matter" but "alive" and thus deserving of respect, dignity, and even love.

The philosophy of spiritual ecology also suggests that we must be careful not to rely solely on scientific answers to address the ecological crisis; we also need to change human behavior. Changing human behavior is a complex ethical challenge. How can our understanding of cosmic consciousness help with this endeavor? I argue that Bucke and Carpenter's conceptualization of cosmic consciousness asserts a clear moral imperative: beyond clime and creed, we are materially bonded by the energies of the cosmos, and our capacity to appreciate this bond is the ultimate source of these creaturely feelings we call "love." Bucke and Carpenter believe that the evolution of the cosmos has a purpose, a purpose that can only be realized (experientially) as the desire to love and be loved. Such a position has clear ethical implications for how we treat each other as well as how we treat the most omnipresent form of "the other," nature. I refer to this ethical system as "the ethics of shared vulnerabilities." By this I mean Bucke and Carpenter were convinced that our shared vulnerability to the vicissitudes of living in a finite world bonds us, and the recognition of this bond upon which "every atom belonging to me as good belongs to you,"[40] not only morally justifies, but encourages us to care for one another, particularly, to care for the oppressed and less fortunate among us. This is an ethics of the heart that privileges relationality over autonomy. Thus, it is not just a matter of putting yourself in the shoes of the other, but about realizing that we share the same shoes by virtue of our communal connection with the cosmos.[41] The significance of deploying this type of relational ethics for spiritual ecology is clear: we are vulnerable to the behavior of nature and nature is more vulnerable than ever to our behavior, thus it only makes sense that we try to cultivate a relationship of mutual flourishing, which is precisely the goal of spiritual ecology. However, an ethical system is only as strong as the metaphysical system upon which it is rooted. I fear that without metaphysical anchoring, the ecological critique of this "ethics of shared vulnerabilities" has no teeth and thus lacks bite. To address how this limitation might be mitigated, I now turn to the philosophy of Alfred North Whitehead.

COSMIC CONSCIOUSNESS AND THE PHILOSOPHY OF ORGANISM

The philosophy of Alfred North Whitehead (1861–1947) can largely be read as a metaphysical critique of modern science's failure to address the totality of our lived experience. In contrast to the mechanistic assumptions that shaped the emergence of the modern scientific worldview, which privileges the study of essential substances, Whitehead's philosophy stresses a much more processual viewpoint that focuses on relationality over substance. Perhaps the essence of his critique is captured most succinctly in his refusal to follow the conventional path blazed by modern science, which aims to bifurcate nature and thereby frame our relationship to the world in dualistic terms. Whitehead believed that this bifurcation of nature distorts our understanding of reality, and thus he posited a processual interpretation instead, a position which he believed was much more attuned to "the nature of things." It is precisely this focus on process and relationality that has inspired certain ecological scholars to declare Whitehead as "one of the patron saints of deep ecology."[42]

The relationship between ecology and Whitehead's philosophy shares a rich and complex history. This is understandable given how well Whitehead's philosophy accords with certain views of ecology. For instance, the core principle of ecology is that everything is interconnected, and this notion of interconnection or relationality is central to Whitehead's thought.[43] According to Whitehead, "we think of ourselves as so intimately entwined in bodily life that man is a complex unit—body and mind. But the body is part of the external word—continuous with it. In fact it is just as much nature as anything else there—a river, a mountain or a cloud."[44] Ecological notions like "web of life," interdependence, and holism find a strong metaphysical base in Whitehead's naturalistic philosophy. But perhaps the most ecologically valuable idea in Whitehead's system of thought is his refusal to separate mind and matter. By exchanging this dualistic framework for a "modified holism,"[45] Whitehead's philosophy provides a coherent and robust argument for the defense of an experiential self-identification with nature. For, as John B. Cobb Jr. and David Ray Griffin point out, "when we have existentially realized that we are continuous with the environment, that the environment is our body, then we will find new styles of life appropriate for that realization."[46] But we can only cultivate new ecological ways of *being-in-world* (based on our self-identification with nature), if the world is designed in such a way to make such a realization possible. Whiteheadian metaphysics asserts that such a view of the world is not only possible, but probable.

The question remains as to how well the ecological significance of Whitehead's metaphysical system aligns with the teachings of cosmic consciousness. Recall that cosmic consciousness refers to the spiritual awareness of one's experiential (and thus embodied) connection to the cosmos. Can Whitehead's metaphysical system admit such an occasion? The short answer is yes. Generally speaking, Whitehead did not address mystical states of consciousness in any great detail, nor did he have a favorable view of reading religious experience in supernatural terms. But the comprehensive nature of his corpus combined with the breadth of his focus, leaves room for a variety of interpretations.

The first thing to consider is that, similar to William James's notion of "radical empiricism," Whitehead was open to expanding our philosophical scope of inquiry to include all forms of experience. For instance, in *Adventures of Ideas* (1933) he writes, "we must appeal to evidence relating to every variety of occasion. Nothing can be omitted, experience drunk and experience sober, experience sleeping and experience waking."[47] Indeed, it would appear that such extraordinary occasions such as those deemed "mystical" would fit such a standard—provided that there is reliable evidence. A comparative overview of the history of mystical experience in both the East and the West would suggest that such evidence does exist. Of course, the question then becomes: by what measure of interpretation do we gauge the reliability of such evidence? If an appeal to supernatural explanations is out of the question, then the naturalistic explanations provided by Bucke and Carpenter, who frame cosmic consciousness as a product of evolution, ought to be considered a legitimate source of evidence according to Whitehead's criteria. From this perspective, we can perhaps interpret cosmic consciousness as a unique occasion that could possibly, in the words of Whitehead, "direct insight into depths as yet unspoken."[48]

Another avenue of approaching a Whiteheadian interpretation of cosmic consciousness is to consider the way he structures the concept of "God." According to Whitehead, we can make a distinction between the Primordial and Consequent natures of God. Briefly put, the Primordial nature of God refers to the ordered ground of the world, while the Consequent nature of

God refers to God's capacity to be impacted by processes in the world: we can experience God, but God can also experience us.[49] It thus follows for certain process thinkers, such as John B. Cobb Jr., that "the affirmation of God's consequent nature is explanatory of some aspect of widespread religious experience."[50]

Perhaps the most fruitful way of constructing a dialogue between the philosophy of organism and the concept of cosmic consciousness is by way of aesthetics. I tend to agree with Matthew Tarnas Segall's insightful analysis of Whitehead in this volume, which outlines the ecological implications of "Whitehead's rejection of the bifurcation of nature in favor of an aesthetic ontology."[51] Beyond viewing this process as a form of play (with which I agree), how can this aesthetic ontology help us to make sense of cosmic consciousness? In his paper, "Cosmic Ecstasy and Process Philosophy" (2005), R. Blair Reynolds suggests that on the surface level Whitehead's conception of aesthetic experience can be interpreted as essentially mystical in the sense that "cosmic consciousness is a major dimension of anything fundamentally aesthetic."[52] But on a deeper level, Reynolds claims that Whitehead's aesthetic ontology defends the notion that "feelings make truth-claims."[53] In other words, feelings, such as those deemed "cosmic," matter, not only as a source of pleasure, but as a way of *knowing* about the world. Thus, if Whitehead's metaphysical system places high value on aesthetic experience and cosmic consciousness is an aesthetic experience, it follows that Whitehead's metaphysical system would value cosmic consciousness. Reynolds succinctly captures the ontological significance of this logic as follows:

> In the context of Whitehead's aesthetic, then, there are no reservations about cosmic consciousness; it is neither a mere cultural epiphenomenon brought on by cruel and unusual sensory deprivation, not a capricious act of grace on God's part; rather, it denotes a fundamental operation throughout the natural order; it represents the most ancient and primal, hence the purest form of experience, by virtue of which the universe is a genuine harmony.[54]

But where would the "ethics of shared vulnerabilities" fit into this aesthetic ontology? Quite clearly I believe in the foundational principle guiding Whitehead's overall metaphysics—namely, creativity. Whitehead believed that creativity is the sacred fabric by which the warp and woof of our existence comes into being and continually guides nature into new forms of becoming. Creativity is the watchword of our existence, and positioning ourselves and the world in harmony with the changes that come with creativity is ultimately an ethical endeavor. From this perspective, the "ethics of shared vulnerabilities" compliments Whitehead's metaphysics by underscoring the ethical challenges that arise when we accept our shared vulnerability to the incessant flux of creativity—for even God cannot escape change! But what is the source of this moral wherewithal to accept change and perhaps even appreciate it? Interestingly, Bucke, Carpenter, *and* Whitehead come to the same conclusion: love. It is the lure of the "Divine Eros" that encourages us to reach new ethical heights of creative harmony with our self, each other, and nature.

CONCLUSION

I began this essay with the suggestion that we need to construct new ecological narratives to counter the anthropocentric ideologies that currently dominate our social and political worlds. Our mass identification with these ideologies has alienated us from nature, which, in turn, has led many of us to believe that we are justified to engage in exploitative practices that threaten the health of the planet. When framed as a principle of spiritual ecology, I argue that the concept of cosmic consciousness can provide a much needed corrective to this dangerous idea that we are destined to master nature.[55] The central teaching of cosmic consciousness is that we are spiritually *and* materially connected to the cosmos and that this connection has ethical implications for how we view and value nature. This cosmic perspective suggests there are alternative ways to conceptualize the relationship between self and nature that do not separate the two or privilege one over the other. Rather, according to Bucke and Carpenter, cosmic consciousness shows that transcendence and immanence, mind and matter, consciousness and energy, soul and body, and self and nature, are mutually inclusive categories that are ontologically coextensive. Such a nondualistic position accords well with the metaphysics of Whitehead's philosophy of organism, which adamantly contests the bifurcation of nature. Moreover, the relational dimension of the "ethics of shared vulnerabilities," which I identify with the "cosmic sense," finds firm metaphysical grounding in Whitehead's philosophy, particularly when we consider his interpretation of God as "the great companion—the fellow sufferer who understands"[56]

There is no question that scientific advancements will continue to radically alter the social landscape of our world. But to what end? It is hard not to marvel at the fact that we can place a man on the moon and yet not solve homelessness. How is it that we have somehow managed to harness the mighty power of the atom, and yet children continue to die from starvation? What are we to make of this strange disproportionation between technological success and ethical failure? This is an important question to address when we consider the future inequalities at stake, once the allure of robotics, artificial intelligence, and the colonization of space, begin to reach a fevered climax in our collective imagination—if they have not already. Today, only a certain economic class can afford to enjoy the pleasures of the latest technologies. Can we imagine a future that will be any different? But, of course, the question of who can or cannot afford the latest designer baby, cybernetic enhancement, or trip to Mars, will be moot if we do not take direct action today to prevent ecological catastrophe. Thus, there is no more urgent challenge confronting the future of life on this planet than that of figuring out how to *live deliberately with nature*. Ultimately, I've tried to demonstrate that there are alternative ways of thinking about our relationship to nature that are still waiting to be discovered—ways of *being-in-the-world* that can perhaps harmonize our scientific aspirations with our spiritual sensibilities. It's possible that a deeper engagement between the ecological teachings of cosmic consciousness and Whitehead's philosophy of organism might very well provide a solid starting point for this discovery.

NINE

Technosophia:

A Cosmohumanist Manifesto

Theo Badashi

INTRODUCTION: OUR MOMENT

As we face the myriad of ecological crises before us, the very fate of our species comes into question.[1] Will humanity survive the impending environmental and political cataclysms? Will our species live to see the emergence of a new human civilization? Will we find the strength and clarity to forge a new human mode of being, one in creative harmony with the planet and universe? Or will we continue on the path we are currently headed, toward the destruction of human civilization and much of the life on Earth?

It is widely discussed amongst diverse circles that our species is in desperate need of a rapid and dramatic transformation. Some scholars and activists suggest that the emergence of a new political and economic system may solve most or all of our major problems. Others believe that the major issues of the world will somehow work themselves out, with the application of more and more advanced technologies, and that technological innovation will be the primary key to our salvation. Though political revolution and technological innovation are indeed necessary for the future survival of our species, I feel that neither force is capable of addressing the magnitude of our current crisis from within their present orientations.

For us to create new social and economic systems requires the participation of human technologies, that is clear. But it is a growing consensus that our present *industrial mode* of technology is far too predatory, destructive, and ecologically oppressive for it to bring about the changes we need. For us to transcend the current techno-economic worldview, and to move into a new paradigm that positions the human within a greater evolutionary context, it is clear to many of us in the New Cosmology movement that we cannot simply adopt new technologies, or make simple adjustments to the current systems we live in. Instead, I believe, we need to introduce an entirely new way of being in the world: a new cosmo-phenomenological worldview that transforms nearly every aspect of human existence, from our fundamental sense of self, to our various modes of being and technological expression. This new worldview, that I have termed *Cosmohumanism*, may just be a vehicle through which such a transformation may occur.

In this chapter, we will explore some of the fundamental patterns and principles underlying this new techno-ecological view of reality. Drawing upon insights from a diverse field of thinkers, and proposing some original ideas of my own, we will examine some of the core questions of cosmology and technology in the hopes of creating a new lens through which to view the evolution of the human species, and the greater evolution of the planet and cosmos as well.

TECHNOLOGY AND THE THREE WORLDVIEWS

Technology, as a process and mode of expression, is an integral aspect of the evolution of the planet and its beings, and has played a fundamental role in the evolution of humanity since before we were human. While possibly all species have their own unique forms of technological expression—as does the Earth itself, I will later argue—in many regards we could consider humans to be a quintessentially technological species, as our very identity is often defined by the symbolic, technological expressions we create.

Though this may be the case, the diversity of human reality has created equally diverse perspectives and relationships with the phenomena we call technology. To begin our discussion, I would like to first clarify what I mean by technology, and then examine some of the major perspectives on technology active today.

TECHNOLOGY DEFINED

In the Socratic tradition of the ancient Greeks, *technology* (from the root words *techne* and *logos*) had often complex and diverse meanings, nearly all of which revolved around the accumulation and application of knowledge for various means. At its core, techne was understood to be *applied knowledge that manifested in both material and nonmaterial forms.* While the word techne carried diverse implications, it was primarily applied to areas related to day-to-day life, and most commonly referred to general crafts and those who practiced them. It is important to note that from its conception, "technology" has always been multivalent, simultaneously encompassing *actors* (the artists, craftspersons, builders, etc.), *processes* (the gathering of means and application of knowledges), *constructs* (the material or intellectual products created), and *use of such constructs* (the functions and applied realities of given technologies). For much of European history technology was conceptualized in similar utilitarian terms. More recently, innovator and technologist W. Brian Arthur further defined technology as: *1) a means to fulfill a human purpose; 2) an assemblage of practices and components; and 3) a collection of devices and engineering practices available to a culture,* founded upon the "capturing and harnessing of natural phenomena."[2]

These foundational insights help us to situate technology within the context of human life, and my use of the word aims to partly embody these definitions. However, if our goal is to understand technology from within a larger evolutionary context, as both ecologist Thomas Berry and technologist Kevin Kelly suggest, then we must understand the archetypal basis of technology and how it manifests in ecological and cosmological terms. By replacing the words "human" with "organism," and "devices/practices" with "constructs," and expanding the definition to include the evolution of nonhuman entities as well, I offer a new cosmologically grounded definition of technology:

> *Technology*: psychic and material constructs created by organisms for the benefit of their survival, actualization, and evolution; and all cultural, ecological and cosmological forces and processes involved in such creations.

In light of this new definition, we can see how the creation of a bird's nest is just as technological as the creation of a human's house, different only in process and degree of complexity. We will return to this cosmological orientation later, but for now it is important to note that the general conception of technology within modern consciousness reflects that of Arthur, as the word increasingly refers to advanced *human* artifacts, technoscientific cultures, and related processes: computers, corporations, and large human-made systems like the Internet. But as our expanded definition suggests, a purely anthropocentric conception of technology only serves to further separate us from both nature and the archetypal existence of realities that could very well be considered "technological." Before we elaborate on this new definition, let us first address the three major technological paradigms that are presently active within modern industrial society, and why a new technological worldview is indeed necessary.

THREE VIEWS OF TECHNOLOGY

In modern industrial culture, I have identified *three dominant philosophical positions* on technology, together forming a diverse spectrum of culture and being. Each position has its own (often interwoven) historical lineages, proponents, values, and ideals. The three dominant technological positions are:

- *The Naturalists*, who primarily emphasize *subjective ecological* technologies—such as psychology, ecology, spirituality, religion, education, and (in many cases) psychedelics—as means of evolving human consciousness to be in greater harmony with the reality of the natural world. Naturalists are essentially *postmodernists* and *ecological phenomenologists* who seek to use technology to explore the relational essence of nature, and who often feel unfavorably about most advanced scientific-materialist technologies.

- *The Technologists,* who primarily emphasize the use of *objective material* human technologies to evolve our species—such as computers, genetic enhancement, and human augmentation. Technologists are essentially *modernists* and *scientific materialists* who view the evolution of life in fundamentally mechanistic, bio-chemical, and genetic terms, and who use technology as the primary means of exploring the physical universe for the ultimate benefit of the human species.

- *The Cosmohumanists*, who seek to synthesize the most generative attributes of these two polarities by utilizing both *subjective ecological* and *objective material* technologies to evolve our species and planet. Cosmohumanists are essentially *integralists*. They seek to understand the significance of both the subjective and objective basis of reality, as well as aim to understand the vast expanse of human potential and the interconnected relationships between humans, the Earth, and our planetary community.

Of these three positions I will be focus primarily upon the distinctions between Technologists and Cosmohumanists.

The Cosmohumanist position is of great significance, I believe, in part because it has the ability to understand the cultural and ontological realities of both Naturalists and Technologists in ways neither position has yet to achieve, and is able to explore and utilize diverse technologies from within each paradigm. Furthermore, Cosmohumanists can ultimately serve as cultural ambassadors between these diverse groups, allowing for greater understanding, empathy, and collaboration to emerge.

An example of a lived Cosmohumanist paradigm may be a person 1) who has a grounded understanding of a psychology, philosophy, and spirituality that is informed by a deep relationship with nature and the living universe; 2) who also has a functional to robust relationship with science and advanced technology; and 3) who may even look to the use of these technologies as a means to explore (and enhance) psychology, spirituality, creativity, and education. In this way, Cosmohumanism aims to be a truly integral orientation toward human existence, cosmic evolution, and technological expression.

It is important to note that these labels serve as fictional markers on a spectrum of consciousness, and rarely exist in pure, clearly demarcated forms. For example, there are some Naturalists who maintain a more subjective-ecological view of the world, but at the same time use advanced technologies—such as ecologists or spiritualists who use the Internet—while also rejecting the dominant *culture* of technology. Similarly, many Technologists may utilize subjective technologies like psychology and psychedelics, while still holding a predominantly scientific-materialist orientation.

The fundamental difference between these two paradigms is that Naturalists often consciously and unconsciously seek to *unify their lives with Earth's living systems* and thus heal the human-nature divide, while Technologists seek to *actualize* the human by developing *mastery over nature* and thus removing humans from the ecologically induced cycles of suffering.

In my view, both objectives hold their own unique attractions, and both provide new potentials for human reality. While Cosmohumanists naturally hold an ecological orientation that resembles that of the Naturalists—a deep desire to come into greater relationship with the Earth and its living communities—the exploration of human potential is also a fundamental draw, as is the scientific exploration of nature and the universe. However, the ways in which we explore nature—both in intent and methodology—are fundamentally different than that of the Technologists. Such alternative modes of exploration have been elaborated upon by thinkers such as Alfred North Whitehead, Thomas Berry, Janine Benyus, Brian Swimme, Richard Tarnas, and others. At its heart, the Cosmohumanist orientation toward nature and science is fundamentally *relational*, where science and technology are seen as tools and methods that allow participants to come into relation with other subjective intelligences, be they other species or entire ecological systems.

TRANSHUMANISM OR COSMOHUMANISM: TWO PATHS FORWARD

Not only do Naturalists and Cosmohumanists have very different views of nature—and humanity's relationship to it—than Technologists, we also have a very different vision of what the future of our species can and should look like. The dominant view within the techno-scientific paradigm of *Transhumanism* is that humans are the rightful masters over nature and will soon transcend both our current biological limitations and present hominid status. Through the accelerated use of some of the most powerful technologies available—and technologies still yet to be developed—humans will come to thoroughly comprehend and manipulate nature as to become capable of bringing the entire evolutionary process (the small fraction within human reach, that is) into our complete and direct control. The ultimate aim of this endeavor, for many in the Transhumanist camp, is the creation of entirely new lineages of posthuman super beings and artificial organisms, intimately interconnected within a global digital infrastructure, at the heart of which lives an omnipotent Artificial Intelligence entity (or entities) monitoring, cataloging, managing, and controlling the entire network. Not only may we one day transcend the limitations of the basic—and they argue *inefficient*—biological bodies and networks nature has thus provided, we appear to be progressing toward the total elimination of all human disease, suffering, and even death. Such is the future of the Transhumanist. Or so they intend.

For humans to successfully move toward this "digital utopia" requires radical advancements in life extension, neural and cognitive technologies, and the further creation of artificial organs; most importantly the synthetic brain. (Some of these technologies are also supported by Cosmohumanists, yet within a different philosophical and ecological context than that of Technologists, a position worth noting.) Regardless of whether or not science achieves its aim of designing a synthetic digital human brain, and then further succeeds at transferring human consciousness from our organic brains into this new digital substrate, the evolutionary trajectories of Transhumanist posthumans may take a variety of different paths.

If we extend current trends in Artificial Intelligence (AI) and Virtual Reality (VR), it is not difficult to image a future synthesis between biological and technological life. In one scenario, large portions of the population may choose—or be forced—to upload their psyches into AI-curated digital atmospheres, while their bodies are either (1) contained in suspended animation in order to extend their physical life spans, mostly discarded and reduced to the most basic functioning components necessary to allow for our brains to still function, or 2) entirely replaced by purely artificial life support systems including new synthetic brains, leaving nothing but a human ghost in a digital machine. In another scenario, Transhumans who choose to stay in physical, non-*Matrix* form will either use technologies—such as genetic engineering, nanotechnology, and artificial augmentation like neural implants and synthetic organs—in order to 1) craft advanced, superhuman cyborg bodies, or (2) choose to transfer/incarnate into completely artificial robot-humanoid bodies, and thus leave behind the organic altogether. Further, some humans may even choose to move between artificial *Matrix* environments and physical cyborg/robot bodies. In the imagined, hyper-real future of the Transhumanists, the possibilities indeed appear limitless.

But what about the rest of the biosphere? Transhumanists will not simply stop at their own radical post-biological transformation, and instead envision the systematic engineering of our entire human and ecological environment: from altering the genetic code of most organisms to the creation of entirely new organisms and biological constructs—a proposed cure for species extinction—even to the full scale "impregnation" of the biosphere with a virtual legion of nanobots whose goal is to render all organic systems accessible to the incarnation and control of omnipotent AI entities.

While these ideas may sound outlandish, impossible, or even hellish, they are nonetheless being actively pursued by some of the most powerful and well-funded technology companies and communities in the world, from Google to DARPA to Facebook. And while this may be science fiction to some, for many of the most ardent proponents of Transhumanism this is exactly the future they envision, or variations within a similar philosophical trajectory. But while these visions appear to be utterly terrifying for many Naturalists, as many do for me as well, I have chosen to take a more holistic approach to these trajectories: rejecting some based upon moral and ecological merits, while entertaining the evolutionary potential of others, a topic we will return to later.

For both Naturalists and Cosmohumanists, however, the future we imagine is very different. Instead of gaining total mastery *over* nature, and transcending our humanity in the process, Cosmohumanists seek to enter into such a profoundly deep relationship *with* nature that our total lived experience and deepest sense of self is fundamentally changed. We do not seek to separate ourselves from nature, but to unify with it in the deepest physical, psychological, and *technological* sense. While Transhumanists aim to explore human-made Artificial Intelligences, Cosmohumanists may aim to engage with the inherent intelligences that generate and govern all livings things: Nature's Intelligence (NI). From this place of deep relationship with life, with other beings, and with the planet itself, we look to the wisdom, intelligence, and design insights of the cosmos in order to create human systems and technologies in harmony with the dynamics of evolution. Through this process a new human identity will emerge, marking the birth of a new form of cosmic-human planetary consciousness. The exploration of this new mode of human being is the prime focus of the next part of this essay.

THE COSMOHUMANIST PARADIGM

Transhumanism and Cosmohumanism offer two very different narratives of a posthumanist future, and I believe that these two worldviews will come to be the dominant paradigms of the twenty-first century and beyond. While I frame the Transhumanist paradigm in relatively negative terms, I personally have a much more nuanced view of it, as it is indeed a widely diverse culture with many important insights to offer. Furthermore, I believe that collaboration between these two paradigms will be absolutely essential, but how successfully they co-evolve together will depend upon how eager each is to empathically engage with one another. While the

Transhumanist cultural identity has emerged over decades and is solidifying before our eyes, the Cosmohumanist cultural identity is still in its early stages of articulation, though its origins and influences can be traced for centuries within both Western and global civilization. For a truly coherent alternative to Transhumanism to emerge, it is necessary that we articulate some of the core values and insights that animate and direct this emerging Cosmohumanist worldview. To do so, I would like to begin by outlining a series of propositions that I feel are at the heart of this new techno-ecological paradigm. I am creating this outline based upon fundamental ideas and positions within a global community of philosophers, scientists, psychologists, ecologists, and spiritualists. I propose four core components that form the foundation of this Cosmohumanist paradigm:

1. A Universe-oriented Cosmology: The Living Universe Story
2. The Validation of All Being(s) and Modes of Being
3. The Emergence of Participatory Teleology
4. A Cosmological View of Technology: Technosophia

COMPONENT ONE: A UNIVERSE-ORIENTED COSMOLOGY

The first fundamental characteristic of Cosmohumanism is marked as a shift away from the current anthropocentric narrative that views the universe as essentially mechanistic, random, and without purpose, to one that is *cosmocentric* and that recognizes the coherence and sacredness of life and the deep intelligence of the universe. To understand how we inherited our current "disenchanted" story of the cosmos, we have to look at the impacts of the Christian worldview and the influence of the Scientific Revolution that emerged in response.

Origins of the Technological Narrative

Regarding the impact of the dominant Christian cosmology on the religious consciousness of Europe, psychologist Edmund Bourne writes:

> For more than a thousand years—up until the early seventeenth century—Earth was widely perceived as the center of the universe, surrounded by a series of concentric spheres that comprised Heaven.... [Everything] that happened in the world was understood to be created and sustained by God or divine providence.

> Humanity had a unique and central importance in this scheme. Not only did human beings inhabit a world that was at the center of the universe, they also occupied a fulcrum point in the vast hierarchy of reality. This hierarchy began with God—its apex—and extended down to the lowliest creatures and plants. Humanity was assigned dominion over the animal and plant kingdoms....
>
> In this hierarchy, ordinary humans were subject to the authority of the church (vested in the pope and bishops), and women and children were subordinate to men. Few questioned the relative order of authority in this hierarchy because it was believed to be the revealed will of God. Everything had its precisely arranged place in the total plan, and obedience to God required not questioning the reality of this cosmic order.[3]

Within this religious society knowledge and authority were ultimately centralized into the hands of the Church and its sanctioned officials. Much of the philosophical and scientific traditions that were practiced by pre-Christian pagan intellectual society had dramatically atrophied and were strategically suppressed under Christian rule. Gradually the religious and spiritual authority of the Church began to extend further and further into realms previously occupied by scientists and philosophers. As a result, most scientists were forced either to conform to the intellectual constraints of the Church, developing their craft underground and in secret, or to face harsh political repercussions, often including exile or death. By the fifteenth century, however, a rising tide of change was forming, brought forth by Copernicus's astronomical observations and the reemergence of the theory of the heliocentric solar system, an idea posited by Greek astronomer Aristarchus more than a thousand years before.

The gravity of this theory was immense. Contrary to the geocentric view held by the Church, the heliocentric model placed the sun at the center of our solar system, and the Earth as one of many celestial bodies revolving around it. With this one powerful idea, humans were essentially cast out from our place at the center of Creation, not only cosmologically but spiritually and psychologically as well. Our very identity as unique creations of God directly and indirectly came into question, along with the Church's scientific and political authority. European consciousness increasingly found itself in the midst of a profound existential crisis. If the Earth was not the center of the universe then what did that mean for humanity? If the Church was not the ultimate authority of Truth, then who or what was? A paradigmatic vacuum had been created, fueling the rise of a new social authority that would ultimately come to take the place of the Church.

The Rise of Modernism and the Fall of God

While prominent scientists, mathematicians, and philosophers—like Francis Bacon and René Descartes—often included God within the emerging secular modernist worldview they were essentially birthing, the ontological basis of this new paradigm no longer required the presence of an omnipotent deity, or of a centralized religious authority. Instead, religious dogma was replaced by democratically verifiable scientific data available to anyone with the right methods and tools. This new mode of inquiry, known as the scientific method, marked a radical democratization of knowledge that simultaneously shifted social and political control away from the Church. It empowered a new generation of free thinkers, enabling amateur and professional scientists to explore nature on their own terms, and thus draw their own conclusions about reality.

Thus the Age of Enlightenment was born, and along with it a new modernist paradigm that placed truth and personal autonomy into the hands of the individual agent. From the early cosmological insights of Copernicus, Kepler, and Galileo, through to the philosophical and scientific propositions of Newton, Descartes, Bacon, and others, and ultimately climaxing in the nineteenth and twentieth centuries with the profound evolutionary insights of Darwin and (in part) Wallace, a new vision of the universe emerged. This vision was based on the existence of fundamental physical laws that could be verified by anyone with the proper tools and knowledge. The scientific revelations that emerged during this period—including evidence that suggested the universe is governed by inherent, seemingly unconscious patterns, and that life itself appeared to be the product of impersonal, mechanical forces—formed the foundation of a radical secular worldview that left little room for the existence of a micro-managing God and even spirituality as a whole.

However, while modernist consciousness saw itself as a natural revolution against the traditional religious consciousness of the time—an identity war that can still be seen within the Dawkins-esque, neo-Darwinian culture of militant atheism—it nonetheless inherited several negative psychological constructs that actively inform its current identity. The two most powerful constructs that directly pertain to our discussion on technology are the beliefs that humans reside at the center of the universe, and that we rightfully have dominion over the Earth and all its systems and beings. While modern scientific culture *consciously* presents a narrative in which humanity is but one small speck within an infinite sea of life and cosmic phenomena, *unconsciously* it acts as if the Earth and entire cosmos are ours to possess, control, exploit, and commodify. We may not be "chosen by God" but we have certainly been chosen by *evolution* to be the true Masters of the Universe.

The Shadow of Progress

The view that the universe is essentially mechanistic, that humans and other organisms are simply organic, gene-propagating machines, and that all of life (even consciousness itself) can be reduced to and understood within purely material terms, forms the foundation of the modern Technologist worldview. When taken to its furthest conclusions, life comes to be viewed through an entirely utilitarian lens, in which organisms and ecosystems are "valued" not on their own subjective realities, but on the functions they serve to other, more "advanced" organisms or systems. It is from this view that common phrases like "food chain" and "Earth's resources" have emerged, and the impacts of this (mostly unconscious) view can be seen in the ceaseless destruction of the Earth's forests and oceans, as well as in the history of species subjugation and forced extinction.

However, this anthropocentric view does not simply apply to nature and nonhuman beings. It has historically manifested as a violent and oppressive form of Euro-American white male supremacy in which all *other* peoples become subject to the Promethean conquest of the white male ego. As such, the history of science and modernism is intimately interwoven with the history of Euro-American colonialism, slavery, genocide, and the objectification and oppression of countless women and queer/transgendered peoples, indigenous peoples, species, and ecosystems. While it can be argued that the pursuit of progress was founded upon some of the most noble of intentions—the understanding of nature, the end of suffering, and the liberation of the individual—the shadow side of progress is impossible to ignore, and indeed must be intentionally addressed and healed if a new sexually, racially, and ecologically just scientific paradigm is to emerge.

Perfect Conditions for Life: The Cosmological Anthropic Principle

As the scientific materialist worldview came to influence much of Western consciousness, a new culture of anti-spiritual atheism and agnosticism rose to guide the ethics and values of modernist industrial society. Central to this was the view that the cosmos was neither sacred nor inherently valuable, and that life on our own planet was likely accidental and therefore meaningless, a view held by many Technologists to this day. However, as the tools of science evolved in their ability to measure and explore the cosmos, many scientists were suddenly faced with unexpected findings that shifted their views on the evolution of life.

Throughout the middle and end of the twentieth century, new information began to emerge suggesting that the vision of a lifeless, accidental—and essentially purposeless—universe needed to be reexamined. Scientists began discovering that the very foundation of the cosmic laws and tendencies that govern the evolution of the universe appeared to be perfectly mathematically tuned for the creation of life and advanced intelligence. Physicists such as Brandon Carter, John D. Barrow, Frank Tipler, and Roger Penrose used these findings to form diverse versions of a radical new proposition, known as the *Cosmological Anthropic Principle*—one version called the *Strong Anthropic Principle*—that posits that *life and intelligence are fundamental components of the evolution of the universe.* Simply stated, the mathematics behind the development of the cosmos are so precise that any major change to any single variable would render the emergence of life impossible. The universe, it appears, is fundamentally orientated to support the perfect conditions that allow stars to give birth to planets, planets to give rise to biospheres, and biospheres to give rise to increasingly complex and intelligent organisms.

While mystics have been making similar propositions for thousands of years—albeit in their own languages reflective of their cultures and traditions—what is culturally significant about this new position is that it has emerged from within the established modern scientific community and is based upon widely accepted scientific theories and observations. While most proponents of the Cosmological Anthropic Principle often continue to frame the universe in material, mechanistic terms, the position that life accidentally emerged within a random, lifeless cosmos is now increasingly difficult to defend.[4]

All of this allows for new philosophical conversations to emerge, with the potential of transcending the purely mechanistic, purposeless view of reality. As we develop a new integral cosmology, we can look to our allies in the modernist scientific community for guidance and insights into the fundamental laws and developmental patterns of the cosmos. But I believe it will be the role of Cosmohumanists to go even further: to bring forth a new cosmological worldview that recognizes the fundamental role of life and consciousness within the evolution of the universe, and grounds those insights into a new post-anthropocentric, *cosmocentric* paradigm. The ultimate aim of this endeavor, I posit, is to create a new cosmology in which the universe *itself* is the central character in this story, and humans are but one small (yet significant) facet of the evolution of a larger *cosmic organism*.

From this new Cosmohumanist position—looking at biological life, and even humanity itself, not as a random "accident" but as a function of the universe's own evolutionary processes—we can now explore a new cosmology that simultaneously humbles and exalts the human. We are humbled in that we are no longer at the center of our own cosmology, but exalted in that we are now characters in a living story that dramatically extends beyond us. The *Universe* is now the Great Subject, and we humans—like all beings—are living aspects of this subject.

In order to expand our own narrative to encompass the larger scope of our cosmic significance, we can begin by examining how the story of the universe has unfolded over time: from the theorized "beginning," when all known matter, energy, light and space is posited to have existed within an infinitely tiny point of origin, through the creation of all the galaxies, stars, planets, and celestial bodies, up through the emergence of life and intelligence on our own home planet. This brings us to the second component, validifying the existence of all aspects of reality.

COMPONENT TWO: THE VALIDATION OF ALL BEING(S)
AND ALL MODES OF BEING

Cosmogenesis: Deep Time and Deep Psyche

In a new Cosmohumanist worldview, our aim is to explore the full spectrum of cosmological phenomena, with the hope of creating a new human reality that validates the universe's own diverse psychic (mental) and material existence(s). To do so we cannot simply examine the physical evolution of the cosmos, nor can we stay tethered to the notion that all of life can be understood in material, biochemical, and genetic terms. The physical sciences have an absolutely critical role to play in the creation of a new cosmic story, but make up only half of this story. The contribution of the other half of the story—exploring the nature of consciousness and the subjective reality of the universe—appears to be the responsibility of philosophers, psychologists, shamans, mystics, and, essentially, *all* beings with subjective experience.

Initially proposed by the French Jesuit paleontologist, theologian, and early integral philosopher Pierre Teilhard de Chardin, the concept of Cosmogenesis can be understood as *the evolutionary phenomenon in which the universe is, and has been, in a continual process of formation since the beginning of time.* Simply stated, Cosmogenesis is the total evolution of the Universe as one singular process of unfoldment.

By using the tools of modern science, in particular, mathematics, and humanity's unprecedentedly powerful telescopes and satellite arrays, scientists have come to understand that the universe as we know it is not the static, materially eternal realm that the ancients once believed it to be. It is an evolving process that has undergone profound and irreversible changes throughout the 13.8 billion years of space-time. This new cosmological understanding of the physical evolution of the universe is widely recognized by modern scientists and Cosmohumanists alike, but where Teilhard and other integral thinkers differ from consensus scientific cosmologists is in the fundamental need to expand this understanding of evolution beyond purely material processes; that is, to include nonmaterial, psychic, and subjective processes as well.

While the physical evolution of the universe may be in a sense more "knowable," through the use of our scientific tools, the psychic evolution of the universe may be of equal, or perhaps greater, significance in regards to the *underlying purpose* of Cosmogenesis. This brings up a deeply heated and long running philosophical debate between three dominant schools of thought around the relationship between matter and consciousness in regards to the evolution of the universe.

The Consciousness-Matter Debate

For countless people throughout history, transpersonal states of consciousness—such as shamanic, trance, dream, and archetypal states—have regularly offered insights into the vastness and ultimate significance of inner-space, suggesting that the importance of consciousness cannot be simply understood through the examination of the material functions and processes of the brain. While philosophers and mystics of both ancient and modern societies emphasize the importance of consciousness as the gateway into spiritual realities, the scientific-materialist worldview has historically denied consciousness of having such significance. Instead, the exploration of virtually all phenomena has been reduced to the material correlations of such experiences: thoughts are less "real" than the signatures they have on our brains; dreams and other transpersonal experiences are merely chemical phenomena with little to no scientific value. In real-world terms, this can be seen in the popular use of prescription drugs—and soon-to-be-available neural implants—designed to impact the electrochemical states of our brains, and thus impact our phenomenological experiences. The scientific conquest to solve human suffering through techno-material means, while ignoring or denying subjective, psychological, and transpersonal solutions and modes of healing, is one of the most clear examples of the limits of the materialist worldview.

Scientific materialists—the lineage most influential in Western culture—generally subscribe to the position that *consciousness arises out of matter* (some scientists and philosophers go so far as to posit that consciousness is a *nonphysical state of matter*), and that consciousness as a phenomena can be reduced to material, electrochemical processes. Consciousness advocates, on the other hand, often hold the position that *consciousness is primary* and that matter is a derivative: that consciousness is at the foundation of being in the universe, and that matter is better understood as a *symbolic-material representation of psychic forces*. Some mystics, including many Hindus and some in Western esoteric traditions, hold the position that matter does not actually exist at all. Everything is consciousness, and the material realm is, in fact, a mental projection, a dream in the mind of the One Cosmic Being.

Between these two polarities lies a third position, well articulated within Buddhist and Hermetic lineages, as well as within the work of Alfred North Whitehead. This position posits that matter and consciousness are mutually co-arising and are actually two different facets of one thing. Matter is real, but is ever imbued with (and reflective of) consciousness. Consciousness is real, but dependent upon matter for its form and the content of its experience. Without a material world (it appears) there would be nothing for consciousness to perceive, nor would there be the material faculties (brains, eyes, nervous systems) to perceive it with. Yet without consciousness, there would be no perceiving in any form, because the basis of consciousness (I argue, cautiously) is that of perception and experience.

For me, the Hermetic axiom *As above so below* comes to mind, positing that mental realities (thoughts and experiences) correspond to material representations (within matter, artifacts, neurological processes, etc.).

Likewise, material realities intimately impact our psychic, subjective experiences. *Consciousness impacts matter, and is continuously impacted by matter.* This further emphasizes the interconnected relationship between the interior subjective and exterior objective basis of reality. My own Cosmohumanist orientation honors the physical and psychic evolution of the universe simultaneously, while *recognizing the primacy of consciousness* as the foundation of subjective experience.

As philosopher and cultural historian Richard Tarnas proposes, a new cosmology must seek to validate the exploration of both the far reaches of the physical universe, as well as the inner recesses of our own psychic, subjective space. We must garner the capacity to move "Deep in and deep out: into the inner depths of the self and our relationship to the Earth, and out into the outer depths of the cosmos."[5] It is this exploration into the inner realms of consciousness—into the archetypal, transpersonal, unconscious, subjective, and intersubjective realms—that places the Integral-Cosmohumanist paradigm into a new ontological position. This new position is simultaneously informed by the culture and insights of the modern scientific community, but transcends it in the ability to explore regions of reality only accessible through subjective inquiry. This brings us to the heart of Component Two, the need to validate all beings and all modes of being in the universe.

The Validation of All Being(s) and Modes of Being

In the established materialist view of reality, validity of existence is most often applied only to that which can be measured in physical terms, and legitimate truth claims made only by those in physically oriented fields of research, such as physicists, biologists, and chemists. While the exploration of material laws and tendencies has lead to the infinite achievements of modern science, such as the harnessing of natural forces for technological use, this narrow view of reality has also created the conditions for wide-scale marginalization—I argue *systemic oppression*—of any form of experience that lies outside of established scientific reality.

From the position that only material reality is worthy of recognition and validity, nearly all forms of subjective experience are reduced to either physical terms (as electro-biochemical-genetic processes), or their significance is disregarded altogether. Emotion and intuition are considered to be "irrational," with little to no utilitarian value. The consciousness of most nonhuman beings, such as frogs, is often considered to be either insignificant (as it is merely an electrochemical process in an organic machine), nonexistent (only humans possess "true" consciousness), or secondary in importance when compared to its physical presence and the environmental function it serves.

If consciousness is either insignificant, nonexistent, or secondary to function, then nearly all beings become subject to purely material forces, to be managed and used by those who have the means to do so. The marginalization of the interior, subjective reality of organisms is one of the single most violent, destructive, imperialistic contributions of the scientific-materialist paradigm. Healing this violence, through the revalidating of subjective experience, both human and nonhuman, must be one of the core aims of a new planetary cosmology.

Though framed in different terms, this appears to be exactly what philosopher and mathematician Alfred North Whitehead had in mind when he set out to heal the chasm between consciousness and matter that materialist science had created. For Whitehead, the reduction of phenomena to abstract terms was a grave disservice to the dynamic nature and significance of subjective experience. While mathematics could certainly be used to describe the physical processes of phenomena, it lent virtually no insights into the nonphysical experiences of such. As Whitehead writes:

> When you understand all about the sun and all about the atmosphere and all about the rotation of the earth, you may still miss the radiance of the sunset.[6]

It is within the subjective experiences of the sun's radiance that we come to know and understand the sun in terms virtually indescribable through abstract, symbolic language, much like a topographical map of a mountain or canyon cannot accurately represent the lived experience of traversing such landscapes. Naturally, this argument extends to all realms of subjective experience, human and otherwise. It was through this revalidating of consciousness that Whitehead imagined a new spiritual-scientific paradigm emerging, one that honored both the subjective and objective realities of existence, and the interplay between them.

Drawing upon Whitehead's mission, the Cosmohumanist position holds that the subjective reality of all beings becomes the foundation for what we perceive as sacred and valid. In this sense, Cosmohumanism is, in part, a lived and democratic *phenomenology* that honors individual perception and experience and holds this subjective reality at the center of all truth claims and notions of validity. The power and potential of an Integral-Cosmohumanist worldview is that it validates all modes of being, both subjective and objective—human and nonhuman—in terms of individual and collective existence.

Using the *All Quadrant All Level* (AQAL) model of analysis, posited by Integral theorist Ken Wilber, we can explore nearly any organism or phenomena in a method that honors the validity and diversity of its being. As we broaden the field of ecology to include the subjective and intersubjective realities of organisms, we can begin to grasp the full complexity of ecological systems in ways that are truly holistic. Organisms are no longer viewed as autonomous organic machines, merely serving the function of propagating genetic code, but can be experienced as individual subjects, in communion with other subjects—a notion ecologist Thomas Berry famously suggested—all of whom have a rich and mysterious interior reality beyond our grasp of knowing. Taking this position requires that we reaccess our views of subjective experience and how we engage with our nonhuman relatives.

If we truly came to accept the subjective validity of all beings on this planet (and beyond), the entire human presence on this Earth would change dramatically. Cosmogenesis asks us to take such a position. By recognizing that humans are both unique in our role and function in the evolution of the universe, yet are only one facet of a larger psychic-material process that transcends and includes the human—"cells of the cosmic organism," as philosopher Spyridon A. Koutroufinis once noted in personal communication—we can begin to extend our empathic experience beyond the human and material realms, to include the full spectrum of consciousness and life.

How then would we choose to live if we recognized the evolutionary significance of the subjective and objective realities of cosmological expression? This question brings us to the third proposed component of the Cosmohumanist paradigm: a new view of the participatory nature of the universe.

COMPONENT THREE: EMERGENCE OF PARTICIPATORY TELEOLOGY

Humans as Earth Keepers and Species Allies

If this new cosmology places all organisms within the context of an evolving Universal Organism, and validates both the interior and exterior manifestations of this organism, then what it means to be participating in this process of Cosmogenesis becomes open to radically new interpretations. The evolutionary process, on this planet and likely countless others, can now be seen as fluid, malleable, and subject to the psychic and material influences of organisms within the biosphere. Cosmogenesis is not something happening *to us*, but a process in which we, and all other beings, are fully *participating*. But what makes humans unique is that our collective actions and decisions impact the planet in ways that transcend any other single species or ecosystem. At the center of the Cosmohumanist moral ethic is the recognition that humans have emerged as a *planetary force,* impacting the evolution—and survival—of countless beings and life systems. In terms of the evolution of our biosphere, such a level of single-species impact appears to be a new planetary phenomenon.

As noted by historian Yuval Noah Harari[7] and others, around 45,000 years ago, humans set out from Africa and headed south to colonize Australia, and, once there, began to initiate a process of exploration and species extinction that has since come to define much of the human migratory experience. Encountering large, primarily herbavoric mammals like the massive marsupial *Diprotodon*—who had been evolving rather peacefully for millions of years with little to no threats from major predators—the early settlers immediately took advantage of the newfound docile protein sources, and within just a few thousand years pushed countless megafauna to the point of extinction. One after another entire species fell victim to this newly emerging hominid predator. Similar patterns occurred on nearly every land mass, with humans arriving at the Americas around 16,000 years ago, and wide scale megafauna extinction shortly following.

Though humans were primarily vegetarian (as grazing was often far more accessible than hunting game) our unique cognitive capacities and technological proclivities—i.e., language and toolmaking—gave us unusual advantages over most species we encountered. That we were pushing species to extinction appeared to be a relative non-issue for our ancestors, up until the rise of the Industrial Revolution, only two hundred years ago, and the full-scale human assault on the planet's biosphere. Now, in the early phase of the largest mass extinction in 65 million years—in which human activity is a known cause of extinction for thousands of species each year—most humans in industrial society are just beginning to realize the truly planetary impact we have. That we have this level of impact is one thing; that we *know* we have this level of impact is another. Not only have humans emerged as the first species—to my knowledge—capable of both exploring and inhabiting nearly every major ecological system on the planet while impacting the evolution of large numbers of species within such systems, humans are the first species to emerge to *realize* our very presence has such a profound planetary impact.

Bleak existential introspection aside, I believe that this moment of self reflection marks not only a significant point in how we view and engage with other species, but it offers a deep-history perspective on the emergence of a genuinely new phase of cosmological development (in terms of known life on this planet, that is). With this single act of realization, humans have now stepped into a new phase of planetary existence, in which we no longer act in the world without knowledge of our impacts. At this stage, the process of evolution moves from unconscious activity into conscious participation. This is not to suggest that all other organisms on this Earth lack self-reflexive awareness, as it is not my place to make such claims. What I am suggesting is that no other organism on this Earth, to my knowledge, has developed the complex cognitive and symbolic capacities necessary in order to comprehend our evolutionary impacts *and* to participate with the fundamental forces of evolution to the extent that humans now have. Further, no other species has yet to move into the position in which it is both self-reflexively conscious of its own existence, and capable of intentionally determining its own evolutionary unfoldment—and indeed the unfoldment of other species and the biosphere as a whole—on a systems-level scale. In this regard the human appears to be unique: *a new mode of planetary existence.*

Because of this, it is essential that humans move beyond our roles as dominators and destroyers into new roles as keepers and protectors of our human and nonhuman families, and the life systems we dearly rely upon. In the act of rising to become the evolutionary directors of Earth, we are immediately humbled in the light of the responsibilities it entails. For most people, it appears, the task we have been given is both unfathomable and unattainable. It is clear that our current mode of techno-industrial exploitation and addiction is no longer sufficient for such a level of responsibility. Instead what is needed is a new human ethic to emerge, one that is profoundly informed by a deep empathic resonance for our entire planetary community. Put simply, for our species to move forward successfully—which includes the preservation of all beings that we can save—humans in industrial society must virtually fall back in love with life, with nature, and with the beauty and mystery that resides in the heart and consciousness of every living thing. Such a revolution is both unprecedented, and absolutely essential, if we wish to survive into the future.

Participatory-Teleology

Originally posited by the Greeks, *teleology* is understood as the progressive movement toward a particular purpose or goal. Aristotle posited that the teleological impulse of an acorn was to grow into an oak tree. Christians (often drawing upon Plotinus and the NeoPlatonists) later argued that the universe, as a singular Being, existed in perfect order, with a conscious God moving the cosmos through a series of stages toward an ultimate reunification of matter and spirit. Unlike more traditional Christian conceptions of a teleological universe, which assume the existence of an all-knowing, intentionally creating God that has "planned" its own unfoldment within every minute instance, *Participatory-Teleology* takes a complimentary but radically different position.

Participatory-Teleology posits that there are indeed cosmological tendencies to move toward greater complexity, but these tendencies are not necessarily the work of a "conscious" deity. Though a deity may or may not be present (indeed a debatable topic depending on one's beliefs and experiences), its intentional intervention ontologically appears unnecessary for the universe to unfold, as the majority of physical processes appear to be governed by unconscious, physical laws. Regardless of whether there is a Supreme Deity, the process of Cosmogenesis unfoldment appears to be more "democratic" than both Greek and Christian theologians initially understood.

Within the human, the universe has moved from unconscious unfoldment to conscious participation. Humans have become co-creators of the evolutionary process: a *consciously participating force* in the universe's own actualization. We are still subject to the cosmic powers that drive the larger patterns of evolution (cosmologist Brian Swimme refers to some of these "powers" as *allurement, centration, differentiation,* and *autopoiesis*), but the ways in which these forces manifest are partly within our realm of conscious influence.

If we view Participatory-Teleology as a central aspect of the process of Cosmogenesis, then this new planetary cosmology places humans into a radically new position of power and responsibility. What this implies is that the human organism—and the physical and nonphysical manifestations we create—is, on one hand, serving a greater evolutionary function that transcends the human sphere, while also changing the entire nature of how this greater evolutionary process unfolds. Our culture is no longer just our own, but part of the Earth's culture as well; our technologies are not just ours but are serving the evolution of the entire Earth community. Naturally, this places a deeper significance upon how we live, the ways in which we engage with other beings, and the diverse impacts of the technologies we create.

It is from this larger cosmological viewpoint that we can now turn our lens upon the phenomena of human and cosmic technological expression. As we proceed, we will take a deeper look at humanity's role in the technological evolution of the planet, the significance of self-reflexive awareness, and deeper insights into the nature of cosmic participation.

COMPONENT FOUR: TECHNOSOPHIA, A NEW COSMOLOGY OF TECHNOLOGY

So far I have presented a series of propositions, drawn from a wide range of established sources, that I believe will help guide our species—especially those of us in industrialized societies—toward a new cosmological relationship with nature and the universe. My fundamental aim in presenting the core propositions of Cosmohumanism is to show how these ideas will ultimately influence human culture, and our technological mode of being in particular. I would now like to present what I believe to be a new *Cosmology of Technology*, one that is informed by both the scientific Technologist paradigm, and the eco-phenomenological Naturalist paradigm. To begin I would like to offer some introductory comments on the nature of technological perspectives, and how a Cosmohumanist might view technology differently than both Technologists and Naturalists.

Technological Perspectives

Naturalists and Technologists occupy two opposing positions that form the foundation of the current technological debate. But while each position holds inherently valid perspectives, they often fail to recognize the flaws within their own respective worldviews. Many Naturalists hold aggressively anti-technological views—including the idea that modern advanced technology is a "wrong turn" in our evolution; or that technology is inherently "unnatural"—thus leading them to disregard the benefits that many technologies may provide, especially in regards to solving climate change and rapidly evolving collective human consciousness. In contrast, most Technologists hold unquestioningly pro-technology views—such as the belief that nearly all major problems can/should be solved with the application of more technology and more human control—without recognizing the negative impacts technology often has on ecosystems, nonhuman entities, and on our deeper collective human cultures and psyches.

Cosmohumanists, on the other hand, might consider ourselves empathetic children of these two worldviews: synthesizing the insights that we feel to be valid and beneficial, while criticizing and rejecting narrow-sighted or destructive tendencies. For example, genetic engineering in itself may or may not be a negative or dangerous technology, though many Naturalists understandably view it as a negative force. What is *actually* dangerous about genetic engineering, I believe, is not necessarily the technology itself, but the conditions of its presence: who controls it, how it is used, and what impacts it has on our species and planet. If technologies like genetic engineering or advanced robotics remain solely in the control of oppressive military-industrial forces (like Monsanto and DARPA), within an ecologically exploitative worldview, Cosmohumanists would naturally oppose such technologies on the basis that they are intended for imperialistic use and greatly threaten the freedom and evolutionary trajectory of people, organisms, and ecosystems.

However, if genetic technologies were explored from within a truly holistic and moral-ecological culture, respectful of the rights and health of all beings involved and considerate of the impacts on the planet as a whole, then Cosmohumanists *might* accept (though not necessarily embrace) its presence and use. While manipulating the genes of an organism in order to patent and control its presence constitutes a clearly undemocratic, non-holistic use of these technologies (as Monsanto aims to control major sections of the world's food supply, for example), genetic technologies like gene therapy may serve our species well in regards to curing diseases and extending human life.

Naturalists have legitimate reasons to be concerned with genetic technologies. I am inherently skeptical and strongly critical of their use. But the genetic manipulation line has already been crossed (with who knows what type of experiments occurring in secret facilities around the world), and while many Naturalists wish to completely reject or disengage with these technologies, Cosmohumanists have a moral responsibility to participate at the core of the conversations that surround them.

Similarly, technologies like robotics, machine automation, and "smart" networking (via the *Internet of Things*) hold profound positive and negative potentials, depending upon the consciousness of the culture or the entity developing them. Automation will soon eliminate a vast number of jobs around the world, which could be a good thing if a Universal Basic Income is implemented along with a radically new, ecologically based—and socially just—economy and education system. If not, automation may lead to even more dramatic economic disparity, while "smart" networking may lead to greater surveillance and the rise of a global techno-fascist state.

Artificial Intelligence, Augmented Reality, and Virtual Reality are three areas I find both incredibly exciting, and powerfully terrifying. These technologies hold unimaginable potentials for our species, and will likely transform nearly every facet of our social, economic, and educational lives. Guided by cultures founded upon empathy, justice, and ecological sensitivity, these technologies can play important and revolutionary roles in our mission to heal the nature-human divide, address the climate crisis, and help unleash a radical new phase in human actualization and potential. Left in the hands of the current techno-capitalist culture, these technologies will likely lead to even greater levels of human isolation, addiction, consumerism, psychological fragmentation, and ecological destruction. In the worst case scenarios, these technologies may come to fully absorb our species within an inescapable *Matrix*-like prison in which oppressive AI comes to subjugate nearly all life on Earth. While it is important to be aware of these dangerous potentials, what is *most* important, I believe, is that a new movement of cosmic visionaries step forward to offer a new narrative of how these technologies can empower the emergence of the ecologically awakened future so many of us imagine.

One area that I see as a positive intersection between Technologist and Cosmohumanist cultures is in the emerging fields of "neurohacking" and "biohacking." Both these terms refer to the intentional exploration of human potential through the use of physical and cognitive training, diet, and experientially enhancing substances like nootropics and psychedelics. Many of the current methods being explored (particularly in the field of nootropics) have shown interesting results in developing greater mental capacities, have been used to increase the neuroplasticity of the brain and increase memory and learning, and have been used to prevent the early onset of deteriorating diseases like Alzheimer's. Furthermore, a new culture of consciousness hackers routinely combine nootropics with intentionally gauged regimen of psychedelics, in the form of microdosing LSD and others. I personally believe that no matter how one finds their way onto the neuro/bio hacking paths—through Technologist or Cosmohumanist channels—the very exploration of consciousness that is often entailed can lead to profoundly expansive and transpersonal states of being, potentially resulting in deeper spiritual and ecological awareness, expanded empathy, and human capacity. As these trends develop we will no doubt witness a maturation of the Cosmohumanist orientation as it continues to form a cosmological narrative that embraces and explores the full spectrum of these diverse states of consciousness and being.

From Unconscious to Conscious Technological Expression

While humans and prehumans have been using technologies for millions of years, their evolutionary significance and overall impact on consciousness, culture, and the living Earth community has gone relatively unexamined until recently. Technology was just "something we did," not necessarily recognized as a force impacting the greater direction of evolution. It is important to note that countless human communities and cultures, throughout history, have understood the power of their technologies to evolve consciousness. That understanding appears to be at the very foundation of human expression, as the artists that projected their visions on to cave walls clearly recognized that their expressions had an impact on them. But again, in the earliest stages of artistic expression, it is difficult to say how reflexive the artists were about the images they were creating. If self-reflexive consciousness was emerging in tandem with the emergence of tools, symbols, and art, then maybe the earliest expressive acts were still very much acts of the unconscious mind, and their impacts on psyche just below cognitive analysis. Maybe it was not until later that creativity crossed over into the full light of reflexive awareness—in tandem with the emergence of rational consciousness?—as the participants didn't just *feel* they were being impacted by the art they were creating and experiencing, but *reflected* upon the fact that it was impacting them.

I would like to suggest that the same emergence of reflexive thought may apply to technology as a whole, and that our own modern Western culture is beginning to examine the evolutionary impacts of our technologies through a new critical lens. Industrialization, the climate crisis, and the hyper complexification of digital communication are but a few of the factors enabling a new techno-reflexive consciousness to emerge. Having now achieved this new form of reflexive awareness, we find ourselves faced with new opportunities to design, create, and explore technologies in ways previous humans had yet to experience. We have awoken to find ourselves imbedded in a participatory universe, more conscious than ever of our creative capabilities, now faced with the task of crafting a new techno-cosmological mode of being founded upon a new sense of intention and purpose.

Humans as a Technological Mode of the Earth

So far we have been focusing upon the technological expression of the human species, but what about other "advanced" beings? If technological complexification is a phenomena that appears to arise within diverse organisms in the universe, as clearly seen on our own planet, then why would similar (or identical) patterns not develop elsewhere? I am not just positing the existence of other advanced sentient beings like humans, which is of course a widely discussed topic, nor am I limiting this view to sentient nonhuman entities on our own planet. (Whales, for example, clearly embody advanced forms of language and culture.) I would like to go further to suggest that technological expression may apply to celestial bodies as well, including our own planet. If this Earth meets nearly every definition of a living organism—exhibiting traits like autopoiesis for example—then is the Earth, as a living being, going through its own process of subjective and technological evolution?

Let us imagine the Earth is, in fact, a subjective being, with its own physical and psychic evolution and, like many beings, it relies upon some mode of technological expression to aid in its actualization process. What then are the technologies of the Earth? To explore such questions we may need to further expand our conceptions of what we consider to be "technology." If we look at the development of the Earth, we might consider the formation of the Earth's material body as one stage in its techno-subjective actualization, and the emergence of life as another stage. The earliest organisms on this planet played the role of creating the biological conditions that allowed for the emergence of more complex organisms: beings with greater psycho-sensory and relational capacities. Did this stage mark the transition from "base survival" to "emotional-relational" modes of Earth actualization? Furthermore, as life evolved on this planet, so did its capacity to express greater forms of creativity and beauty, as seen in the emergence of colorful plants, insects, and eventually birds, for example. With every new evolutionary iteration, the physical and psychic modes of the Earth continued to complexify, until a new form of cognitive awareness emerged within our early human ancestors, possibly within many other prehominid or nonhominid species as well.

What makes humans unique in this process is that we have attained a particularly complex degree of self-reflexive symbolic consciousness and psycho-spiritual-emotional capacity, modes that had yet to emerge within other beings on the planet, to this extent. As Teilhard de Chardin, Thomas Berry, and Brian Swimme collectively posit, it may be this unique form of consciousness that defines the human and offers insights into what purpose and function the human mode of Earth is here to serve. To use Teilhard-Swimme's language, humans may be the eyes, ears, and feeling heart of the cosmos—at least an earthy expression of such. This suggests that the human may have emerged in order to allow the Earth (and essentially the universe as well) to experience more dynamic emotional, creative, psycho-spiritual, and even technological states.

When we look at the evolution of the human within the larger context of Earth's own evolution—with special emphasis on the creative ways that humans express ourselves, including our impulse to create art, build complex structures and systems, and our incredibly rich modes of emotional and intellectual being—interesting hypotheses can be made. If the evolution of early, less complex organisms correlated with the early stages of Earth's own actualization process, then the emergence of the human may signify a *new* stage of planetary and cosmic actualization: the emergence of a complex, psycho-spiritual and techno-symbolic mode of being. In this sense, humans, like all other organisms and bio-geological processes, may be considered a *living technology of the planet*. In this view, we can further expand our definition of technology to include living beings and larger cosmological processes.

Furthermore, we can even go so far as to speculate that the Earth itself is coming to a new level of awareness of how to use the human mode of technological expression in more "wise," "mature," and conscious ways. Just as humans undergo a process of exploratory "trial and error" with all of our technologies, so too the Earth may be exploring the depths, limits, and capacities of its own human technology. As we are becoming conscious of our own technological mode of being, so too may the Earth be becoming conscious of its own human-technological nature as well.

What, then, does it mean to be a techno-symbolic mode of the Earth, and what is required of us now that we recognize ourselves as such? Well, for one, we no longer have the luxury or privilege of making technological decisions purely on behalf of the human, without a deep, integral consideration of the evolution of other organisms and of the planet itself. We now face the paradigm-shifting realization that our mode of technological being is actually not "ours" at all, but is in fact a larger cosmological phenomena being expressed through us, as it is expressed through countless other beings. Our technologies, therefore, do not really *belong* to us, but are a fundamental manifestation of the Earth's own actualization process. To live in full conscious participation with this techno-symbolic mode of being suggests that we must develop a new sense of identity, a new relationship with the planet and Earth community, and a new understanding of the evolutionary significance of the technologies we create. We will explore these ideas more as we continue.

Limits of Industrial Consciousness

The current form of human techno-industrial consciousness deeply lacks any real consideration of how our technologies actually impact the subtle forces of the planet and the psycho-physiological well-being of the Earth community. In spite of how profoundly capable industrial consciousness appears to be in regards to its creative capacities, there is little to no *actual intention* behind the creations of most consumer technologies, outside of purely short-term economic gains. Furthermore, most technologies that emerge from this mode are fundamentally violent in nature, in regards to their overall presence and ecological impacts. The very ontological basis of this industrial mode makes these conditions inevitable, leading to the continuous infliction of profound generational traumas upon most members of human and Earth society. Even the nature of violence itself is rationalized: a necessary byproduct of human progress. This has led to the creation of a global culture fully dedicated to the actualization of the human ego, at the conscious and unconscious expense of nearly all other life.

For example, a corporate marketing culture that seeks to exploit the subtle weaknesses and sexual-survival impulses of humans is very far removed from the actual traumatic impacts it is having on the beings it seeks to influence. Many brilliant, good-hearted people wake up each day with the goal of creating exciting and enticing products and services, in the hopes of creating economic abundance for themselves, their companies, and their families. What they are unaware of is that they are often participating in a predatory cycle of imperialistic dominance that is wounding the mental, physical, and ecological health of their "customers." As long as the imperialistic shadow side of techno-industrial consciousness remains unrecognized, our dominant human culture will continue to perpetuate violence and trauma upon the planet. But as we look into the future, I believe that this cycle of violence can come to an end.

Soon we will see the emergence of a new techno-economic system founded upon a phenomenological relationship with the sacred and subjective reality of the Earth, a system fully conscious of its presence and impacts, and participating in symbiotic alignment with all life. We do not necessarily need to transcend symbolic-currency economic systems, at least not yet. Instead, we have a new opportunity to activate a deeper Earth consciousness within the techno-economic culture, so that it plays out *its desired role*—to bring forth new innovation, to alleviate suffering in the world, and to raise living standards for all—in a way that expands upon the power, significance, and evolutionary responsibility it presently serves. From this generative perspective, eco-preneurs can fully participate in the healing of the human species and the greater Earth community. On our present course, this reality might not come into being before irreparable damage occurs. Indeed, it might already be too late, as we are now witnessing the greatest loss of species and ecological communities in the past several million years, at the hands of human-industrial predation. But, if enough of us move quickly, and help to cultivate this rapidly growing community of cosmo-humans, we may just have a chance to shift the experiential reality of technological society as a whole. I believe this shift is possible.

Cosmomimicry: Aligning with the Psyche of the Cosmos

For us to emerge as truly conscious participants of Earth's actualization process, we must cultivate an entirely new level of conscious awareness in regards to the evolutionary significance of human culture and technology. This is the primary aim of Technosophia: to develop a new conscious, cosmo-phenomenological understanding of virtually every aspect of our technological mode of being and expression. Our ultimate goal, therefore, is to develop an integral understanding of the Earth's actualization process, so that we, as humans, can participate with this process in the most evolutionarily significant ways. This includes a new understanding of the karmic and evolutionary implications behind the technologies we create, including: 1) an awareness of the initial *intentions* underlying the technological creation process; 2) the ecological *design* of each technology; 3) the *creation* process itself, including the *methods* and *context* of each stage, such as the *means* in which we procure material sustenance, the *conditions* in which technologies are produced, and the ecological *impacts* of production; and 4) the *introduction* and *implementation* of new technologies through to the ways in which we continue to *cultivate* their powers and potentials. By developing a conscious awareness of every stage and phase of our technological processes, we have the opportunity to create technologies that are genuinely of the Good, in that the very basis of our technologies reflect the underlying generative forces of the universe, the actualization process of the planet, and work to actualize the greatest potentials of all beings on Earth and beyond.

Using the universe and Earth as our guide, I am suggesting a new mode of technological awareness that I term *Cosmomimicry*. Cosmomimicry is a further expansion upon Janine Benyus's concept of *Biomimicry*, a design philosophy that looks to the biosphere of the planet as a guide in the development of new technologies. While the application of Biomimicry has profound implications for the design of our technologies, and is an important foundational position, the diverse ways in which it is often applied suggests that its focus is, in many regards, still too human-centric. It is simply not enough to design cars, robots, and solar panels to reflect the inherent patterns of the biosphere, though this is an important aspect for sure. For a truly radical shift in human technology to occur, we need to transcend the impulse to apply biological insights to purely material expressions, within a predominantly economic context. We need to go several steps further, toward the creation of a new technological orientation that reframes our entire sense of self, by looking to the universe as our guide in the creation of not just new material technologies, but new psycho-spiritual technologies as well.

With Cosmomimicry, we aim to understand the universe as a great Cosmic Subject, with the intention of understanding the fundamental forces and subjective realities of this organism: how it creates new physical modes, and the interior consciousness behind its creation process. By doing so, we aim to transcend the impulse to understand nature purely in order to create physical technologies for human use. By exploring the subjective *and* objective reality of the universe, we can come to align ourselves with the psychic and material processes of this Being, and look to its archetypal patterning as the ultimate model for how we can create a new techno-cosmological existence for our species.

We can begin by asking questions like: How can we work with the forces of *centration* and *allurement* to create new educational models that help to inspire students and educators to explore the deepest depths of the universe and their own psyches? How can we create regenerative economic systems that transcend the oppression of organisms, and that positively contribute to the greater evolution of all beings? How can we create political systems that are not human-centric, but reflect the sovereignty and sacredness of nonhuman citizens as well? And most importantly, how can we create a new philosophical worldview that allows for a new human techno-cosmological identity to emerge? If our aim is to properly explore these questions, from a truly integral position beyond the current material and economic modes of thinking, I believe we need to move beyond speculative theory, toward direct *phenomenological* apprehension of the subjective reality of the universe.

Technosophia and Cosmohumanist Phenomenology

For us to develop a new ontological orientation toward our human mode of being requires that we transcend the traditional scientific barriers around our ways of knowing and experiencing the world. The primary ontological difference between the dominant Western scientific-materialist paradigm and that of Cosmohumanism is that of the *accessibility of knowledge and insights*. While modern Technologists and materialists most often believe that knowledge and information can be accessed through purely rational means (through mathematics and scientific observation), Cosmohumanists, like Naturalists, believe that valid knowledge and information can be accessed through subjective, experiential, intuitive means as well. By entering into the subjective reality of nature and the cosmos we can come to experience the archetypal and transpersonal dimensions of the universe, and of technology itself.

By moving deeply into our own subjective experiences, and by entering into realms that we may call *shamanic* and *transpersonal*, Cosmohumanists seek to fully unify our phenomenological reality with the subjective reality of the cosmos, in deep relationship with its psychic structures and generative forces. Our aim essentially becomes that of attuning ourselves so acutely with the underlying intelligences of the universe that they experientially guide us, educate us, and show us how we can symbiotically participate with the universe's own creative processes. Thomas Berry, in his deeper reflections on the role humans are to play as a dynamic mode of the Earth, teaches us that the way we come into contact with the subjective reality of the universe is by entering into what may be considered deep post- and pre-rational shamanic states, on a regular basis, and for long periods of time. Only by "moving out" of our limited human minds can we come to experience the mind of the higher intelligence(s) generating this reality. By placing oneself in the presence of the natural world, and by entering into transpersonal and meditative states, the human egoic mind can temporarily subside, and the foundational awareness of the human can come to perceive the underlying, archetypal reality generating the material and non-material phenomenon of nature.

In my own life, I have had direct experience of the states Berry may have alluded to, as have countless others throughout history. For me such states have been experienced in a variety of modes: in the presence of nature (most often for elongated periods of time); in states of deep meditation; in states of ecstatic expression; in the presence of potent beauty; and most frequently through the use of entheogenic substances, plant medicines, and psychedelics. What I have experienced is that when one enters into these transpersonal, holotropic states, how ever such an entrance occurs, the underlying evolutionary patterns and forces that generate this reality are often revealed, and the deeper impact and significance of human activity—and the activity of other beings and the Earth itself—is made perceptible. Such states are now becoming common amongst a growing population on this planet, a sign of the emergence of what Berry called a *new shamanic mode of human reality*.

The shift I am describing has indeed been articulated or alluded to—in various forms—by philosophers and mystics throughout history in traditions such as Hinduism and Buddhism, in the work of philosophers such as Plotinus, Hegel, Aurobindo, Teilhard, Whitehead, Tarnas, Swimme and others, and in premodern practices within Indigenous cultures around the planet. Within the Romantic and Transcendentalist lineages of the West, we see aspects of this shift in consciousness described in the writings of the poet Walt Whitman, and in the writings of his contemporaries R.M. Bucke and Edward Carpenter. While Whitman offered beautiful, poetic, and often sensual windows through which to peer into the deeper interpenetrations of human and nature, it is in the work of Bucke and Carpenter that we see this process articulated in more philosophically tangible forms, notably in their shared use of the term *cosmic consciousness* to describe the stage of psychic development in which human awareness moves beyond the anthropocentric, conceptual ego-mind, and into a "higher" stage of eco-phenomenological unity consciousness. It is at this stage that the sacredness and spiritual magnitude of nature and the universe are not simply conceptualized, but profoundly experienced, thus altering one's fundamental orientation toward life so that one may come to live from a place of deep reverence and connection to nature and all living beings.[8]

Toward a New Cosmological Society

Through the development of a deeper phenomenological relationship with the Psyche of the Cosmos, we can begin to explore how to apply these insights toward every facet of human society. As we come to imbue a new cosmo-phenomenological consciousness into our major human systems—by creating the conditions in which planetary evolution is not simply a theorized external process, but a generative realm that can be subjectively experienced—we may witness the global emergence of an entirely new human-Earth community. What will it look like, and how will we insure that it brings about true planetary actualization, justice and freedom? Again, recognizing the Earth as our primary guide, we may come to understand how it is that the planet maintains its own autopoeisis, and seek to align our own patterns and systems with its underlying impulses. As we can begin to speculate, this path will dramatically diverge from the current mode of human economics, politics, and education.

Economics: Every community of organisms on this Earth appears to exist within complex symbiotic relationships with their neighboring beings and ecological environments. Humans too must develop such relationships, starting with the very sources of our ecological sustenance. A symbiotic economic system could be founded upon the basis of bioregional relationships between all organisms living within a particular ecosystem. As we may come to experience ecosystems as superorganisms—and individuals and communities within this organism functioning as modes of its own being—cultivating the health of the superorganism becomes our primary ecological and economic motive. This is not to suggest that individual autonomy is not to be respected, however, as the health and actualization of individual organisms within each bioregion is absolutely essential for the maintenance of the health of the larger entity. What I am suggesting is that the desires of any individual cannot threaten the overall health of the community if the community is to survive and thrive. To achieve this, all impulses toward *self* actualization could be harmonized within a larger context of *collective* and *ecological* actualization. This implies that the inherent rights of *all beings* are to be fiercely upheld, so that all creatures may successfully pursue their own actualization.

In the human sphere, this will, in part, translate into economic terms. The human-mammalian impulses toward pleasure and the desire for new experiences can be understood within the context of evolutionary unfoldment—as a manifestation of the process of *allurement*—and, instead of denying or distorting these impulses, a new economy can seek to develop evolutionarily significant ways to channel these energies. I believe it will be the cultural responsibility of active Cosmohumanists to help create healthy economic alternatives to many of our current modes of consumer behavior. The cultural "hunger" drive to consume material products can—and ultimately *must*—be recalibrated and channeled toward new forms of expression that promote the actualization of the individual and community, in ecologically sensitive ways. I believe this will be accompanied by a shift away from a dominant materially-consumptive economy, toward more creative forms of psycho-spiritual entertainment and education, creating an entirely new service-experience economy. The creation of new Technosophic technologies will likely play an instrumental role in this recalibration process as well, as the karmic, ecological, and evolutionary factors involved in our economic decisions will be considered within a greater cosmological context. Since we will of course still rely upon material consumption as an important aspect of our economy, the way in which we consume goods could shift toward the *Cradle-to-Cradle* design philosophies as proposed by William McDonough, Michael Braungart, and others. Products could be reframed within holistic, closed-cycle, nutrient-based systems, in which the creation and consumption of each product becomes regenerative and ecologically healing.

Finally, new "resource based" economies founded on ecologically responsible sustenance distribution, in which the primary basic needs of all Earth citizens are freely met, could replace our current scarcity-extraction-consolidation economic model. With the implementation of a *Universal Basic Income*, such a society may be made possible. And with the rapid evolution of the global economic system—in which factors like resource scarcity, carbon pollution, and the automation of most manual production could soon make our present economic models completely obsolete—a new sustenance distribution system will not be simply preferred but absolutely essential for our future human survival.

Politics: As our economies shift toward a more bioregional orientation, so too may our political systems. If more Cosmohumanists choose to engage with the dominant technological culture in the West and beyond, and begin to participate with some of the core insights of the Technosophic orientation, we may bring about the emergence of a new technological culture consciously dedicated to the democratic distribution of advanced technologies. Given that Technosophic technologies are inherently regenerative, and cosmologically oriented, we could then have the chance of actually solving many of the larger planetary crises we face, including issues of ecological and political instability, poverty, and material sustenance. Once the basic issues of sustenance are met, for example, we can then transition away from the paradigm of scarcity, toward that of abundance. This process may literally render our major political systems obsolete, and prime them for a total reimagination of their functional role and overall existence. Though I cannot speculate as to what a new political system will look like, I believe that if we continue to look to the Earth and universe as our guide, the vision of a New Earth Democracy will one day emerge so clearly, within such a powerful diversity of beings, that the path toward its creation will become apparent.

What I will speculate upon, however, are some of the core characteristics that I believe may be present in a new cosmo-political system. First of all, for a new system to be truly cosmological, that is for it to work in harmony with the planet and universe, it must in some sense be democratic, in that it must respect the inherent rights of all living beings, ecosystems, and the planet itself. Since humans are acting as the primary facilitators of Earth's symbolic mode of consciousness, it will naturally be the responsibility of humans to serve as legal representatives on behalf of nonhuman citizens. This will require a new Planetary Bill of Rights to emerge, and an entirely new legal mode of being. If the major issues of sustenance scarcity may be ultimately solved through the creation of Technosophic technologies, the entire military basis of every government on Earth will necessarily shift away from the present aims of "security" (which essentially translates to "resource security"), toward more peaceful and significant modes, such as helping climate refugees and victims of ecological crises, as well as refocusing military discipline toward the peaceful exploration of space. The emergence of new economic and political systems will allow for a truly profound healing to occur within the warrior consciousness of our planet: a healing process absolutely essential if we are to help our military siblings move out of their current survival and predatory modes of consciousness and into new forms of empowered identity.

On a federal and global level, governments could come to serve as facilitators and stewards for regional and global cultural relations to be propagated amongst diverse interconnected communities, and could serve the primary role of maintaining a completely transparent, global legal infrastructure. Eventually, once humanity has moved significantly down the path toward healing the traumas of imperialist consciousness, governments could one day act as Earth emissaries within a larger galactic and cosmic community of beings in the universe: ecologically informed ambassadors of our own Earth community. Though these speculations reach far into the future, I believe that such a future may be ours to create.

Education: As we shift into new economic and political modes of being, our educational orientation will move away from that of "training" students to serve as mere workers and consumers—economic cogs in an industrial machine—toward the goal of actualizing our greatest human and planetary potentials. As this Great Turning continues, the role of education will take on a new, and truly profound, evolutionary significance. Imagine an educational system, as Thomas Berry suggests, centered around the exploration of humanity's place in the greater unfolding of the universe. The Universe Story will be taught as *our* story: the foundation of our cosmic heritage. Students will learn about the exciting 13.8 billion year saga that created every phenomena in the universe, up through the birth of our own galaxy, solar system, and planet, through to the emergence of life, and the infinite beings on this Earth that we call our terrestrial family. All accessible modes of life on this planet will be explored, and students will come to appreciate the infinite diversity of our human and planetary community.

Students could grow up falling in love with all of life, excited to learn about the infinite dimensions of the planet and cosmos, and eager to explore new frontiers of our evolutionary existence. Explorations into science, the arts, and humanities, may come to be informed by a new cosmological sense of self and purpose, with student curriculum centered around how each of these fields apply to their own evolution as cosmic beings. The passion and intelligence of these new cosmological humans may be channeled toward finding new solutions to the larger planetary crises we face, and toward the exploration of entirely new forms of human consciousness and culture.

Informed by the profound insights of scientific research, and imbued with new forms of subjective-phenomenological understanding, students will grow to be truly Integral agents in the world and universe. Imbued with a new sense of cosmological identity, and given the means to explore the full capacities of our creative potential, our new educational systems could serve as the cultural conduits through which a new global Renaissance could emerge, helping to bring forth a new era of human culture, and a new phase of planetary actualization. This is the ultimate aim of the Cosmohumanist worldview: to cultivate a new human species reflective of the vast and beautiful reality of Cosmogenesis.

CONCLUSION: THE EMERGING COSMOHUMANIST CULTURE

As we come to the end of our discussion, I would like to offer a few words in regards to the sources and magnitude of the ideas we have explored. First of all, it is essential for me to state that the ideas I have presented are ultimately not mine—though I may have offered new language in some areas—but in fact represent the consciousness of a growing community of beings on this planet. Furthermore, these ideas do not *belong* to this community either, though they have emerged through this particular human lens. In fact, they may not be "human" ideas at all, but insights and reflections of the planet itself, phenomenologically experienced through one of its various modes of awareness, expressed in a techno-symbolic form enabling the greater actualization of its own being.

With this in mind, I would like to reaffirm my view that although many of these ideas and propositions feel as though they exist in some far off distant place—the future perhaps—more and more of us on this planet are coming to experience them *at this time*. The sheer scope and diversity of humans now exploring such a cosmological reality is an indication, to me, that we have already entered into the early phases of a new planetary stage of development. Perhaps we are further along than we realize, as the seeds of this new cosmological experience may have been planted billions of years ago.

Endnotes

INTRODUCTION

1. We owe particular thanks to Matthew T. Segall for carefully reading the final version.

CHAPTER ONE: DEACON

1. Significant parts of this essay have been borrowed from Terrence W. Deacon, "On Human (Symbolic) Nature: How the Word Became Flesh," in *Embodiment in Evolution and Culture*, ed. G. Etzelmüller and Christian Tewes (Heidelberg: Mohr Siebeck, 2016), 129–49.
2. Terrence W. Deacon, *The Symbolic Species: The Co-evolution of Language and the Brain* (New York: W.W. Norton, 1997).
3. C.S. Peirce, *Collected Papers of Charles Sander Pierce. Vol. II Elements of Logic*, ed. C. Hartshorne and P. Weiss (Cambridge, MA: Harvard University Press, 1931).
4. Deacon, *The Symbolic Species*; Deacon, "Universal Grammar and Semiotic Constraints," in *Language Evolution*, ed. M. Christiansen and S. Kirby (Oxford: Oxford University Press, 2003), 111–39.
5. Terrence W. Deacon, *The Symbolic Species*; Deacon, *Incomplete Nature: How Mind Emerged from Matter* (New York: W. W. Norton, 2012).
6. Terrence W. Deacon, "The Aesthetic Faculty," in *The Artful Mind*, ed. M. Turner and S. Zeki (Oxford: Oxford University Press, 2006), 21–53.

CHAPTER TWO: SEGALL

1. Alfred North Whitehead, *Religion in the Making* (Cambridge: Cambridge University Press, 1927), 4–5.
2. Alex Rosenberg, "How Physics Fakes Design, and Makes Things Difficult for Theism," paper presented at the Greer-Heard Point-Counterpoint Forum in Faith and Culture at the New Orleans Baptist Theological Seminary on February 22, 2014, accessed January 5, 2016, https://www.youtube.com/watch?v=VImtigQ-eIA.
3. Jacques Monod, *Chance and Necessity: An Essay on the Natural Philosophy of Modern Biology* (New York: Vintage Books, 1972), 172.
4. Alfred North Whitehead, *Process and Reality*, corrected edition, ed. David Ray Griffin and Donald W. Sherburne (New York: Free Press, 1978), 23.
5. Terrence W. Deacon, "The Trouble with Memes (and What to Do about It)," in *The Semiotic Review of Books* 10, no. 3 (1999), accessed January 5, 2016, http://projects.chass.utoronto.ca/semiotics/srb/10-3edit.html.
6. Daniel Dennett, "Wild and Domesticated Religions: How the Machinery of Religion Evolved," paper presented at the Santa Fe Institute on March 16, 2010, accessed January 5, 2016, https://youtube/Qo4V7PsX4qU.
7. Alfred North Whitehead, *The Concept of Nature* (Cambridge: Cambridge University Press, 1919).
8. Alfred North Whitehead, *The Function of Reason* (Princeton, NJ: Princeton University Press, 1929), 61.
9. Whitehead, *The Function of Reason*, 69.
10. Whitehead, *Religion in the Making*, 74.
11. Whitehead, *The Function of Reason*, 11.
12. Whitehead, *The Function of Reason*, 21.
13. Whitehead, *Religion in the Making*, 31.
14. Darwin, *On the Origin of Species* (New York: P. F. Collier & Son, 1909), 262, quoted in Robert N. Bellah, *Religion in Human Evolution* (Cambridge, MA: Belknap Press, 2011), 208.
15. Bellah, *Religion in Human Evolution*, 85.

16. Whitehead, *Religion in the Making*, 3.
17. Whitehead, *Religion in the Making*, 5.
18. Jason Kelly, "Shared Vulnerabilities: Cosmic Consciousness and the Philosophy of Organism," chap. 8 in this book.
19. Johan Huizinga, *Homo Ludens* (Boston: Beacon Press, 1950), 17–18.
20. Huizinga, *Homo Ludens*, 3.
21. Whitehead, *The Function of Reason*, 8.
22. Whitehead, *Religion in the Making*, 10.
23. As Joseph Campbell put it in the opening lines of *The Hero With a Thousand Faces*, these myths may be "the secret opening through which the inexhaustible energies of the cosmos pour into human cultural manifestation." See Campbell, *The Hero With a Thousand Faces*, 3rd ed. (Novato, CA: New World Library, [1949] 2008).
24. See Marshall Sahlins, *Stone Age Economics* (London: Routledge, [1972]) 2004.
25. Buber's term was "I-Thou"; Bellahs uses "I-You."
26. Bellah, *Religion in Human Evolution*, 104.
27. Bellah, *Religion in Human Evolution*, 82.
28. Whitehead, *Religion in the Making*, 128.
29. Alfred North Whitehead, *Modes of Thought* (New York: The Free Press, 1968), 111.
30. Whitehead, *Process and Reality*, 167.
31. Whitehead, *Religion in the Making*, 91–92.
32. Huizinga, *Homo Ludens*, 15.
33. Whitehead, *Religion in the Making*, 140, 143.
34. Alfred North Whitehead, *Adventures of Ideas* (New York: MacMillan, 1969), 213; *Process and Reality*, 343.

For Further Reading

On the ideological sources of the "selfish gene" approach to biological evolution: Bruno Latour. "How to Make Sure Gaia Is Not a God of Totality, with Special Attention to Toby Tyrrell's book On Gaia." Written for the Rio de Janeiro meeting "The Thousand Names of Gaia," September 2014.

On the geochemical inevitability of the emergence of life on earth (life is no accident): James Trefil, Harold J. Morowitz, and Eric Smith. "The Origin of Life: A Case Is Made for the Descent of Electrons." *American Scientist* (Volume 97), 2009.

On the importance of love in biological evolution: Humberto Maturana Romesin and Gerda Verden-Zoller. *Origins of Humanness in the Biology of Love.* Imprint Academic, 2009.

CHAPTER THREE: PIKARSKI

1. My thanks to Spyridon A. Koutroufinis and Terrence W. Deacon for their valuable comments on my ideas during the summer of 2015. Further thanks are dedicated to Claire Manuel and Jeanyne B. Slettom for their vital help with editing this chapter.
2. Brandon Flowers, Dave Keuning, Mark Stoermer, and Ronnie Vannucci, *Human* (New York: The Island Def Jam Music Group, 2008), written and performed by the band *The Killers.*
3. Plato, *Alcibiades* (London: William Heinemann, 1964), 199–201.
4. Jean-Baptiste le Rond d'Alembert, *Preliminary Discourse to the Encyclopedia of Diderot* (Chicago: University of Chicago Press, 1995), http://hdl.handle.net/2027/spo.did2222.0001.083.
5. Max Scheler, *Die Stellung des Menschen im Kosmos* (Darmstadt, Germany: Reichl, 1928), 13–14.
6. Ernst Cassirer, *An Essay on Man* (New York: Doubleday, 1944), 40.
7. Alfred North Whitehead, *Process and Reality: An Essay in Cosmology,* corrected edition, ed. David Ray Griffin and Donald W. Sherburne (New York: The Free Press, 1978), 3.
8. Thomas Suddendorf, *The Gap: The Science of What Separates Us from Other Animals* (New York: Basic Books, 2013), 215–30.
9. Suddendorf, *The Gap*, 216.
10. See Spyridon A. Koutroufinis, chap. 5 in this volume.
11. Alex Gomez-Marin points out the important relation between *durée* and intuition in his chapter (chap. 4 in this volume)—which is a perfect companion to mine.
12. Henri Bergson, *Creative Evolution* (New York: Dover, 1998), 151–76.
13. Cassirer, *An Essay on Man*, 41–44.

14. Terrence W. Deacon, *The Symbolic Species* (New York: W. W. Norton, 1998).
15. Often, sustainability is defined as an interdisciplinary approach to solve intertemporal and global problems. It means satisfying the global needs of today without risking the needs of future generations. Thus, to reflect upon sustainability, intertemporal distance is required. But, of course, this is only one side of the story, and maybe the intellect is not the only capacity that has to enable this work. Hence, not only intelligent distance is needed, but also a closeness which emerges from that distance. With Bergson, I prefer to approach this problem using the concept of intuition given in Gomez-Marin's chapter in this volume.
16. It is possible that laughing about someone or something is tendentially more intellectual than laughing with someone. While the first requires more intellectual distance, the second requires closeness, which is more anticipated with instinct and empathy, or with intuition!
17. I am referring to an important German distinction between the words *"erfahren"* and *"erleben,"* which are both translated as *"experiencing."* While *Erfahrung* is dedicated to the more conscious and objectivated (quantified) experience guided by the intellect, *Erleben* is more dedicated to the subjectively experienced life qualities which we more often associate with feelings.
18. Bergson, *Creative Evolution*, 176.
19. Scott Barry Kaufman, Carolyn Gregoire, *Wired to Create* (New York: Perigee, 2015), 64–76.
20. The so-called conflict between science and art is one discourse which only occurs if the intellect is overestimating itself. Bergson's intuition is one way to fulfill Whitehead's claim of a meaningful interweaving of objective knowledge and subjective experience. Only then is an interdisciplinary and coherent interpretation framework for *every* experience possible.
21. Michel Foucault, "The Ethic of Care for the Self as a Practice of Freedom" in *Philosophy & Social Criticism* (Sage Journals, 1987), 129.
22. Michel Foucault, *The History of Sexuality, Volume 1—An Introduction* (New York: Pantheon Books, 1987), 92–102.
23. Michel Foucault, "The Subject and Power," in *Critical Inquiry* 8, no. 4 (Chicago: University of Chicago Press, 1982), 777–95.

24. Michel Foucault, *Discipline and Punishment* (New York: Vintage Books, 1995), 135–94.
25. Michel Foucault, *The Hermeneutics of the Subject* (New York: Palgrave MacMillan, 2005), 183–84, 85–86.
26. Foucault, *The Hermeneutics of the Subject*, 133.
27. Foucault, *The Hermeneutics of the Subject*, 130–31.
28. Foucault, *The Hermeneutics of the Subject*, 94.
29. Foucault, *The Hermeneutics of the Subject*, 235–37.
30. Foucault, *The Hermeneutics of the Subject*, 252.
31. Alfred North Whitehead, *Modes of Thought* (New York: Palgrave MacMillan, 1988), 76.
32. Whitehead, *The Function of Reason* (Princeton, NJ: Princeton University Press, 1928), 2.

CHAPTER FOUR: GOMEZ-MARIN

1. I thank Spyridon A. Koutroufinis, Laura Nuño de la Rosa, Rod Hemsell, Joana Rigato, Juan Arnau, José Gomes Pinto, and Johanna Häusler for insightful comments on this manuscript.
2. In order to propose a precise stance to address the problem of creativity in contemporary evolutionary biology, we found it necessary to rescue and re-enunciate Bergson's notion of *multiplicity* as *heterogeneous continuity*, surpassing the contradictions that both gradualism and punctuated equilibrium share, and accommodating both continuity and discontinuity in evolution. Čapek's warning comes to mind: "Logicians will probably never like the Bergsonian terms 'qualitative multiplicity' or 'heterogeneous continuity'" ("Immediate and Mediate Memory," *Process Studies* 7, no. 2 [Summer 1977]: 90–96.) More than that, Bergson's thinking invites us to dwell in *durée*, which entails a different kind of intellectual effort that allows for *intuition*. One may rightly note that this very same essay could have been written a century ago, and that biology has progressed much since then (so much, some may claim, that it does not even need philosophy at all). Yet, a century after the publication of *Creative Evolution*, Bergson's pioneering insights remain relevant for science in general, and for biology in particular (not to mention philosophy in these times of dominance by the analytical style). Bergson's views also remain mainly neglected—a

dismissal (tainted with disdain) out of the inability (tarnished with unwillingness) to grasp them. Now that a revival of process thinking seems to be taking place with a greater force in the philosophy of biology (and in biology itself), it is timely and important to go back to the source. Process thinking can and must revise the habits and blind spots of the intellect. When discussing evolution we must be very careful of *what our intellect makes of it*.

3. It is quite ironic that both variation and selection can be traced back precisely to the two types of forces that vitalism puts forth: one that complexifies (as an intrinsic power of life), and another that tames (due to the circumstances). Of course, the orthodox evolutionary view adopted them for their own agenda.

4. Herbert Spencer Jennings saw in the behavior of "lower" organisms the manifestation of the same method that Lloydd Morgan had identified in "higher" organisms: trial-and-error as a universal heuristic spanning both in ontogeny and evolution. The same principle is at work for children learning and for bacterial chemotaxis. To see the evolution of species as bacterial navigation, and the converse, is certainly powerful and appealing to the intellect since one can then conceive how anything may not know where to go but still get there. And so, once that act of explanation is accomplished in the mind of the scientist, it is assumed that the phenomenon to be explained follows the same algorithm. The final result is that the bacterium, the child, and evolution all proceed up the gradient both "as if" they wanted to get there and also "as if" they knew how to get there, but, in reality, they do not really know nor want. Yet, ironically, the scientist's "as if" becomes the bacterium's "is," and not the other way around.

5. "For, though the variation must reach a certain importance and a certain generality in order to give rise to a new species, *it is being produced every moment, continuously and insensibly*, in every living being. And it is evident that even the sudden 'mutations' which we now hear of are possible only if a process of incubation, or rather of maturing, is going on throughout a series of generations that do not seem to change. In this sense it might be said of life, as of consciousness, that *at every moment it is creating something*" (Bergson, *Creative Evolution* [New York: Dover, 1998], 30; italics added).

6. Evolution is commonly thought to include Neo-Darwinism but also, and despite their commendable efforts, the movement called The Third Way of Evolution.

7. One may dare to say that the commitment not to accept the truly new reflects what Whitehead called "the philosophy of an epoch," which, actually, embodies a range of systems during the twentieth century that were at war each other, while sharing their inability to conceive novelty in its strong sense. In his words:

> When you are criticizing the philosophy of an epoch, do not chiefly direct your attention to those intellectual positions which its exponents feel it necessary explicitly to defend. There will be some fundamental assumptions which adherents to all the variant systems within the epoch unconsciously presuppose. Such assumptions appear so obvious that people do not know what they are assuming because no other way of putting things has ever occurred to them. With these assumptions a certain limited number of types of philosophic systems are possible, and this group of systems constitutes the philosophy of the epoch (Alfred North Whitehead, *Science and the Modern World* [Cambridge: Cambridge University Press, 1953], 61).

8. "The more we study the nature of time, the more we shall comprehend that duration means invention, the creation of forms, the continual elaboration of the absolutely new" (Bergson, *Creative Evolution*, 16).

9. The Bergsonian proposal can be deemed as originating from a kind of "epistemological pessimism." Indeed, Bergson claims that life does not lend itself to the thinking of the intellect. The intellect finds itself at home with matter and space. Intuition is the faculty that is able to grasp life and be in *duration*. One can then claim the opposite: Bergson's epistemology is genuinely positive and full of force.

10. This applies not only to the world "out there," but to our own memory:

> My memory is there, which conveys something of the past into the present. My mental state, as it advances on the road of time, is *continually swelling with the duration which it accumulates*: it goes on increasing—*rolling upon itself*, as a snowball on the snow" (Bergson, *Creative Evolution*, 8; italics added).

In fact, Bergson's progression from *Time and Free Will* (1889) to *Matter and Memory* (1896), to *Creative Evolution* (1907) carries the same intuition from our inner experience to the brain to evolution; a conjoint theory of life and of knowledge:

unity and multiplicity are only views of my personality taken by an understanding that directs its categories at me; I enter neither into one nor into the other nor into both at once, although both, united, may give a fair imitation of the mutual interpenetration and continuity that I find at the base of my own self. *Such is my inner life, and such also is life in general* (Bergson, *Creative Evolution*, 220; italics added).

11. Bergson points out that the real problem of our theorizing about evolution is the source and cause of variations. He comments on the insufficiency of Darwinism:

> Comme je viens de le dire, Samuel Butler a montré que le darwinisme se trompait en prenant la concurrence vitale et la sélection naturelle pour des principes que suffiraient à expliquer l'évolution des espèces, alors que *ces deux principes peuvent à la rigueur rendre compte de la survivance de telles ou telles variations, mais non pas de l'apparition de ces variations elles-mêmes, qu'il faudra dès lors attribuer au simple hasard*. On alléguera que c'était l'évidence même; et, par le fait, ce jugement porté sur le darwinisme est celui de l'immense majorité des biologistes. Encore fallait-il s'en apercevoir, et il est possible (…) que Butler ait été le premier à faire cette constatation, à la formuler clairement, à mettre en lumière l'impossibilité de porter les variations au compte du simple hasard. (…) *la véritable question, pour la biologie et la philosophie évolutionnistes, est de déterminer les causes de variation* (Bergson, *Mélanges* [Paris: Presses Universitaires de France, 1972], 1523; italics added).

12. "Our mind, which seeks solid bases of operation, has as its principal function, in the ordinary course of life, to imagine states and things. Now and then it takes quasi-instantaneous views of the undivided mobility of the real. It thus obtains sensations and ideas. *By that means it substitutes for the continuous the discontinuous, for mobility stability*, for the tendency in process of change it substitutes fixed points which mark a direction of change and tendency" (Bergson, *Key Writings*, ed. Keith Pearson and John Mullarkey [London: Continuum, 2002] 274; italics added).

13. "The real whole might well be, we conceive, an indivisible continuity. The systems we cut out within it would, properly speaking, not then be parts at all; they would be partial views of the whole" (Bergson,

Creative Evolution, 32).

14. Vladimir Jankélévitch, *Henri Bergson* (Durham, NH: Duke University Press, 2015), 169.
15. "Quelle est cett intuition: Si la philosophie n'a pas peut en donner la formule, ce n'est pas nous qui y réussiron. Mais ce que nous arriverons à ressaisir et à fixer: c'est une certaine imagem intermédiaire entre la simplicité de l'intuition concrete et la complexité des abstractions qui la traduisent, image fuyante et évanouissante, qui hante, inapercue peut-être, l'esprit philosophique, que le sit comme son ombre à traves les tour et detours de sa pensée, et qui, si elle n'est pas l'intuition meme, s'en rapproche beaucoup plus que l'expression conceptuelle nécessairement symbolique, à laquelle l'intuition doit recourir por fournir des 'explication'" (Bergson, *Laughter: An Essay on the Meaning of the Comic*, trans. Cloudesley Bereton and Fred Rothwell [Project Gutenberg e-book, 2003], 3).
16. Karin Stephen, *The Misuse of Mind* (New York: Harcourt Brace, 1922), 36.
17. One could argue that these are also functions of the intellect, and that it may also produce novelty in its way; that it can also be truly creative. In fact, by means of words, my thinking is attempting to be creative in revealing its own tendencies.
18. One could say, à la McKenna (who is, in turn, paraphrasing Whitehead) that the evolutionary process is, ultimately, the creative struggle between habit and novelty.
19. Bergson, *Key Writings*, 60, italics added.
20. Bergson, *Key Writings*, 260.
21. Bergson, *Key Writings*, 246.
22. Bergson, *Key Writings*, 209.
23. Bergson, *Mélanges*, 434, italics added.
24. Spyridon A. Koutroufinis, chap. 5 in this volume.
25. "[A]ction is discontinuous, like every pulsation of life; discontinuous, therefore, is knowledge" (Bergson, *Creative Evolution*, 260).
26. "This amounts to saying that there is no essential difference between passing from one state to another and persisting in the same state. If the state which 'remains the same' is more varied than we think, on the other hand the passing from one state to another resembles,

more than we imagine, a single state being prolonged; the transition is continuous. But, just because we close our eyes to the unceasing variation of every psychical state, we are obliged, *when the change has become so considerable as to force itself on our attention*, to speak as if a new state were placed alongside the previous one. Of this new state we assume that it remains unvarying in its turn, and so on endlessly" (Bergson, *Creative Evolution*, 9; italics added).

27. "The apparent discontinuity of the psychical life is then due to our attention being fixed on it by a series of separate acts: actually there is only a gentle slope; but in following the broken line of our acts of attention, we think we perceive separate steps. True, our psychic life is full of the unforeseen. A thousand incidents arise, which seem to be cut off from those which precede them, and to be disconnected from those which follow. Discontinuous though they appear, however, in point of fact they stand out against the continuity of a background on which they are designed, and to which indeed they owe the intervals that separate them; they are the beats of the drum which break forth here and there in the symphony" (Bergson, *Creative Evolution*, 9).

28. Heterogeneous discontinuity is perhaps a more common view (no connection between species extinction due to driving cars, slaughtering baby seals, and bombing in Iraq or Syria, except on a highly abstract level of analysis).

29. Bergson, *Creative Evolution*, 293.

30. One may say that Bergson is actually thinking of this; that his intellect, like Whitehead's, is able to conceive transformation of essence. Anyhow, the point is that they are not following the natural movement of abstract thinking but transcending it, and so evolution is seen for what it is: heterogeneous continuity. In this sense, thought is not circumscribed to the rules and habits of the intellect.

31. Bergson, *Creative Evolution*, 141.

32. "The evolution of life as a whole, from its humblest origins to its highest forms, inasmuch as this evolution constitutes, through the unity and continuity of the animated matter which supports it, a single indivisible history" (Bergson, *Creative Evolution*, 36).

33. "*The truth is that this continuity cannot be thought by the intellect while it follows its natural movement.* It implies at once the multiplicity of elements and the interpenetration of all by all, two conditions that can hardly be reconciled in the field in which our industry, and

consequently our intellect, is engaged. Just as we separate in space, we fix in time. *The intellect is not made to think evolution, in the proper sense of the word—that is to say, the continuity of a change that is pure mobility*" (Bergson, *Creative Evolution*, 141; italics added).

34. "The aim of this education is to harmonize my senses with each other, to restore between their data a continuity which has been broken by the discontinuity of the needs of my body, in short, to reconstruct, as nearly as may be, the whole of the material object" (Bergson, *Key Writings*, 106).

35. Bergson, *Key Writings*, 107.

36. "Regarded in what constitutes its true essence, namely, as a transition from species to species, life is a continually growing action" (Bergson, *Creative Evolution*, 113).

37. Henri Bergson, *The Two Sources of Morality and Religion*, trans. R. Ashley Audra and Cloudesley Brereton (London: Macmillan, 2014), 96; Auro e-Books (The Internet Archive).

38. Bergson, *Creative Evolution*, 241.

39. Discussing Boex-Borel's work *Le pluralisme; essai sur la discontinuité et l'hétérogénéité des phénomènes*, Bergson says:

> Prétend-on nous faire assister a un passage graduel de l'homogène à l'hétérogène, ou nous faire entrevoir un aboutissement final de l'hétérogène à l'hómogène? Veut-on, par un effort d'imagination, reconstruire idéalement le complexe avec le simple ou le simple avec le complexe? Dans tout les cas, on se heurte à une impossibilité (Bergson, *Mélanges*, 794).

40. Bergson, *Creative Evolution*, 134, italics added.

41. "It would have seen that if, in order to simplify the work and also to facilitate the co-operation, things are first reduced to a few categories, or ideas, translatable into words, each of these ideas stands for a stationary property or state culled from some stage or other in the process of becoming; the real is mobile, or rather movement itself, and we perceive only continuities of change; but to have any action on the real, and especially to perform the constructive task which is the natural object of human intelligence, we must contrive to have halts here and there, just as we wait for a momentary slowing down or standing still before firing at a moving target. But these halts, each of which is really the simultaneousness of two or more movements

and not, as it seems to be, a suppression of movement, these qualities which are but snapshots of change, become in our eyes the real and essential, precisely because they are what concerns our action on things. *Rest then becomes for us something anterior and superior to movement, motion being regarded only as agitation with a view to a standing still. Thus immutability is rated higher than mutability, which implies a deficiency, a lack*, a quest of the unchanging form. Nay more, it is by this gap between the point where a thing is and the point where it should be, where it aspires to be, that movement and change will be defined and even measured. On this showing, duration becomes a debasement of being, time a deprivation of eternity" (Bergson, *The Two Sources of Morality and Religion*, 204; italics added).

42. "The first rule about creativity is that if you rush, you lose it. Like cooking a soup or making love, creativity's highest preoccupation is not efficiency. But we want to cut to the chase and get as much stuff done as quickly as possible . . . Yet, the creative process (precisely because it is a process) just takes time" (Gomez-Marin, "A Portrait of the Scientist as a Young Artist: Or, What Neuroscience Can Learn From Dance," *SciArt Magazine* Vol 33 [2018], https://www.sciartmagazine.com/a-portrait-of-the-scientist-as-a-young-artist-or-what-neuroscience-can-learn-from-dance.html [*Things I Learned from an Artist About Science*, Gomez-Marin, unpublished]).

43. "And that is what intelligence expresses by saying that thus only it arrives at distinctness and clearness. It must, therefore, in order to think itself clearly and distinctly, perceive itself under the form of discontinuity" (Bergson, *Creative Evolution*, 139).

44. Bergson, *Laughter*, 29.

45. "There is not 'ceteris paribus' in biology" (Alex Gomez-Marin & Asif A Ghazanfar, "The Life of Behavior," *Neuron* 104, No.1 [2019]: 25–36, https://doi.org/10.1016/j.neuron.2019.09.017).

46. "The evolution of the living being, like that of the embryo, implies a continual recording of duration, a persistence of the past in the present, and so an appearance, at least, of organic memory." (Bergson, *Creative Evolution*, 23.)

47. "The aim is not to effect another Platonism of the real, as in Kant's system," he contends, *"but rather to enable thought to re-establish contact with continuity and mobility."* (Bergson, *Key Writings*, 36; italics added.)

48. That is, The inherent resistance in our acceptance that one can go beyond the intellect without negating it.
49. "An attention to life, sufficiently powerful and sufficiently separated from all practical interest, would thus include in an undivided present the entire past history of the conscious person—not as instantaneity, not like a cluster of simultaneous parts, but as something continually present which would also be something continually moving" (Bergson, *Key Writings*, 262).
50. "Common sense represents the endeavour of a mind continually adapting itself anew and changing ideas when it changes objects. It is the mobility of the intelligence conforming exactly to the mobility of things. It is the moving continuity of our attention to life" (Bergson, *Laughter*, 57).
51. Jakob von Uexküll, *Theoretical Biology*, trans. Doris Livingston Mackinnon (New York: Harcourt Brace, 1926), 70.
52. Uexküll, *Theoretical Biology*, 72.
53. Bergson, *Creative Evolution*, 22.
54. David Lapoujade, *Potencias del tiempo: versiones de Bergson* (Buenos Aires: Editorial Cactus, 2011), 92.
55. Bergson, *Mélanges*, 429.
56. There has been a renewed interest in process philosophy in biology. Having read what could be considered as the primary sources, I remain puzzled at the "mainstream" recent attempts to re-erect a process philosophy with no mention of Whitehead whatsoever, and with an intellectual style kin to analytical philosophy which, in my view, precisely betrays the essence of process thinking.
57. Whitehead, *Process and Reality*, corrected edition, ed. David Ray Griffin and Donald W. Sherburne (New York: The Free Press, 1978), 24.
58. Whitehead, *Process and Reality*, 29.
59. "[W]hich, at any rate, makes succession, or continuity of interpenetration in time, irreducible to a mere instantaneous juxtaposition in space" (Bergson, *Creative Evolution*, 289).
60. Bergson, *Key Writings*, 223.
61. Lawlor and Moulard, "Henri Bergson," *Stanford Encyclopedia of Philosophy*.
62. Biology too often treats life as a great puppet. Why should the dead

be the template to explain the living, and not the other way around? How could repetition be the engine for creativity? Why reorganization in space is supposed to give rise to transformation in time?

63. Bergson, *Creative Evolution*, 6, italics added.

64. "In the thrust,...effects succeed their cause, which 'produces' them—in the literal sense of the term—by pushing them along. Such is the impulse of a shock of an efficient or efferent cause. But all of Bergson's dialectic consists, precisely, in showing that basically, the same is the case for 'final' causality—in which it is the effect that precedes" (Jankélévitch, *Henri Bergson*, 55).

65. "The coherence, which the system seeks to preserve, is the discovery that the process, or concrescence, of anyone actual entity involves the other actual entities among its components. In this way the obvious solidarity of the world receives its explanation" (Whitehead, *Process and Reality*, 7).

66. "In vain, therefore, does life evolve before our eyes as a continuous creation of unforeseeable form: the idea always persists that form, unforeseeability and continuity are mere appearance—the outward reflection of our own ignorance. What is presented to the senses as a continuous history would break up, we are told, into a series of successive states" (Bergson, *Creative Evolution*, 32).

67. Bergson, *Key Writings*, 148, italics added.

68. Jankélévitch, *Bergson*, 17.

69. "[T]he world the mathematician deals with is a world that dies and is reborn at every instant—the world which Descartes was thinking of when he spoke of continued creation. (Bergson, *Creative Evolution*, 25).

70. "Continuity of change, preservation of the past in the present, real duration—the living being seems, then, to share these attributes with consciousness. Can we go further and say that life, like conscious activity, is invention, is unceasing creation?" (Bergson, *Creative Evolution*, 30).

71. "We are at ease only in the discontinuous, in the immobile, in the dead. The intellect is characterized by a natural inability to comprehend life" (Bergson, *Creative Evolution*, 143).

72. "Bergson ne s'est pas proposé de nous donner de la relation causale une analyse logique; il se place à un point de vue psychologique" (Bergson, *Mélanges*, 439).

73. "La conscience met dans la matière continue une discontinuité: l'être raisonnable cherchera à rétablir la continuité, en vertu du sentiment profond qu'il a de la continuité originelle" (Bergson, *Mélanges*, 441).
74. Bergson, *Creative Evolution*, 21.
75. Spyridon A. Koutroufinis, chap. 5 in this volume.
76. "We claim, on the contrary, that the spontaneity of life is manifested by a continual creation of new forms succeeding others" (Bergson, *Creative Evolution*, 78).
77. "Now, a mystic society, embracing all humanity and moving, animated by a common will, towards the continually renewed creation of a more complete humanity, is no more possible of realization in the future than was the existence in the past of human societies functioning automatically and similar to animal societies" (Bergson, *The Two Sources of Morality and Religion*, 70).
78. Bergson, *Creative Evolution*, 29.

CHAPTER FIVE: KOUTROUFINIS

1. I would like to warmly thank Jason James Kelly and Matthew T. Segall for useful remarks on the early drafts of this paper and for carefully reading the final version.
2. Jakob von Uexküll, *Umwelt und Innenwelt der Tiere* (Berlin: Springer, 1909), 117, 196, 249, 252; Uexküll, *Theoretische Biologie* (Frankfurt/Main: Suhrkamp, 1973 reprint), *Theoretische Biologie*, 2nd ed. (Berlin: Springer, 1928), 320.
3. Jakob von Uexküll, *Theoretical Biology* (New York: Harcourt, Brace, 1926), translation of *Theoretische Biologie*, 1st ed. (Berlin: Gebrüder Paetel, 1920), 127; Thure von Uexküll, "A teoria da *Umwelt* de Jakob von Uexküll," *Revista Galáxia* 7 (2004): 19–48.
4. Jakob von Uexküll, *A Foray into the Worlds of Animals and Humans* (Minneapolis: University of Minnesota Press, 2010), 150.
5. Uexküll, *Theoretical Biology*, 78, 97–99.
6. Uexküll, *Theoretical Biology*, 93.
7. Uexküll, *Theoretical Biology*, 98.
8. Uexküll, *Theoretical Biology*, 103.
9. Uexküll, *Theoretical Biology*, 78.

10. Uexküll, *Theoretical Biology*, 15, 16.
11. Uexküll, *Theoretical Biology*, 126–77.
12. Uexküll, *Theoretische Biologie*, 62.
13. Uexküll, *Theoretische Biologie*, 50, 63, 95, 96, 115, 127, 156.
14. Immanuel Kant, *Critique of Pure Reason* (Cambridge: Cambridge University Press, 1998), B 129–32.
15. Kant, *Critique of Pure Reason*, B 134–35.
16. Kant, *Critique of Pure Reason*, B 130.
17. Kant, *Critique of Pure Reason*, B 137.
18. A core idea of Whitehead's metaphysics is that actual occasions are processes that constitute themselves through internal relations to their environment. Whitehead calls these relations "prehensions."
19. Another core idea of Whitehead is that the self-creation of a process or actual occasion is the growing together of its prehensions (see note 18) to a new unity. Actual occasions are short-lived flashes of existence. After the completion of their self-creation, they exist in space for only a short time during which they can be prehended. Whitehead describes the process, i.e., the new actual occasion that arises from the integration of the prehensions as *concrescence*, from the Latin verb "concresco" meaning "growing together." Thus, the actual occasions are *synthetic* acts of self-constitution.
20. *Vernunft* is *Verstand* guided by principles. *Vernunft* also has a moral component.
21. Kant, *Critique of Pure Reason*, B 76.
22. Kant, *Critique of Pure Reason*, B 132.
23. Spyridon A. Koutroufinis, and Arthur Araujo, "Uexküll, Whitehead, Peirce: Rethinking the Concept of 'Umwelt' from a Process Philosophical Perspective" (forthcoming).
24. A basic premise of Whitehead's metaphysics is the assumption that actual occasions are inseparable *physical-mental bipolar unities*. He conceives of the actual occasions as "processes of experience," (Alfred North Whitehead, *Adventures of Ideas* [New York: Free Press, 1967], 197), which he calls subjects of their own experienced immediacy (Alfred North Whitehead, *Process and Reality, An Essay in Cosmology*, corrected edition, ed. David Ray Griffin and Donald W. Sherburne [New York: Free Press, 1979], 25). Actual occasions are physical-mental

unities. They are processes endowed with subjectivity that are always related to and can also generate things that exist physically in space-time. Whitehead explains that the term "mental" is much more comprehensive than "conscious," as only very few mental phenomena can be classified as possessing consciousness. Nearly all actual occasions are merely *proto-mental* events and as such they are not conscious. Different processes are configurations of widely variable types and may exhibit any number of grades of consciousness, including a complete lack thereof, depending on their complexity. But all processes are complexes of experience. Thus, the idea of experience plays a much greater role in Whitehead's concept of a subject than does the idea of consciousness.

25. Uexküll, *Theoretical Biology*, 53.
26. Ernst Cassirer, *An Essay on Man: An Introduction to a Philosophy of Human Culture* (New Haven: Yale University Press, 1944), 41.
27. Uexküll, *Theoretical Biology*, 126.
28. Cassirer, *An Essay on Man*, 42.
29. Cassirer, *An Essay on Man*, 43.
30. Cassirer, *An Essay on Man*, 43.
31. Cassirer, *An Essay on Man*, 43.
32. Cassirer, *An Essay on Man*, 44.
33. Cassirer, *An Essay on Man*, 48.
34. Cassirer, *An Essay on Man*, 48.
35. Terrence W. Deacon, *The Symbolic Species: The Co-Evolution of Language and the Brain* (New York, London: Norton, 1997).
36. Cassirer, *An Essay on Man*, 52.
37. Cassirer, *An Essay on Man*, 54.
38. Cassirer, *An Essay on Man*, 56.
39. Cassirer, *An Essay on Man*, 59.
40. See Deacon, *The Symbolic Species,* and his chapter in this book.
41. For the role of sexual selection in human evolution see chaps. 19 and 20 of *The Descent of Man* (Charles Darwin, *The Descent of Man, and Selection in Relation to Sex* [Princeton, NJ: Princeton University Press, 1981]).

42. Cassirer, *An Essay on Man*, 64–65.
43. Cassirer, *An Essay on Man*, 66.
44. Henri Bergson, *Time and Free Will* (London: George Allen & Unwin, 1950), 95.
45. Bergson, *Time and Free Will*, 78–79; italics added.
46. Bergson, *Time and Free Will*, 84.
47. We are allowed to add apples with apples but not apples with oranges unless we add fruits.
48. Bergson, *Time and Free Will*, 76.
49. Bergson, *Time and Free Will*, 89.
50. Bergson, *Time and Free Will*, 96.
51. Cassirer, *An Essay on Man*, 72.
52. Cassirer, *An Essay on Man*, 74.
53. Cassirer, *An Essay on Man*, 75.
54. Henri Bergson, *Matter and Memory* (London: George Allen & Unwin, New York: Macmillan, 1929), 22.
55. See Chapter II of Bergson, *Time and Free Will*.
56. Bergson, *Time and Free Will*, 98.
57. Bergson, *Time and Free Will*, 100–101, italics added.
58. Bergson, *Time and Free Will*, 99.
59. Bergson, *Time and Free Will*, 98.
60. Bergson, *Time and Free Will*, 97.
61. Bergson, *Time and Free Will*, 97 [addition by Koutroufinis].
62. Bergson, *Time and Free Will*, 98.
63. Bergson, *Time and Free Will*, 97.
64. Bergson, *Time and Free Will*, 96, italics added.
65. Henri Bergson, *Creative Evolution* (London: Macmillan, 1922), 142–95.
66. Bergson, *Creative Evolution*, 141, 149–50.
67. Cassirer, *An Essay on Man*, 78.
68. Cassirer, *An Essay on Man*, 78–79.
69. I owe this idea to René Pikarski. See his chapter in this volume.

70. Heinrich Rickert, *Die Grenzen der naturwissenschaftlichen Begriffsbildung* (Tübingen: Mohr, 1929).
71. Edmund Husserl, *The Crisis of European Sciences and Transcendental Phenomenology* (Evanston IL: Northwestern University Press, 1970), 123–35.
72. Uexküll, *Theoretische Biologie*, 21.
73. Whitehead, *Process and Reality*, 7.
74. Immanuel Kant, *Critique of Judgment* (Indianapolis: Hackett Publishing Company, 1987), § 65.
75. Hans Jonas, *The Phenomenon of Life. Toward a Philosophical Biology*, (Evanston, IL: Northwestern University Press, 2001) and "Evolution und Freiheit," in *Philosophische Untersuchungen und metaphysische Vermutungen* (Frankfurt am Main: Suhrkamp, 1994), 11–33.
76. Whitehead, *Process and Reality*.
77. Barbara Muraca, *Denken im Grenzgebiet: Prozessphilosophische Grundlagen einer Theorie starker Nachhaltigkeit* (Freiburg, Munich: Alber, 2010), 42.
78. Muraca, *Denken im Grenzgebiet*, 37–39, 46.
79. Herman Daly, *Beyond Growth: The Economics of Sustainable Development* (Boston: Beacon Press, 1996), 51; Muraca, *Denken im Grenzgebiet*, 45–52.
80. Muraca, *Denken im Grenzgebiet*, 173–81, 243–47.
81. Bergson, *Creative Evolution*, 191.
82. Bergson, *Creative Evolution*, 191.
83. Bergson, *Creative Evolution*, 192, italics added.
84. Bergson, *Creative Evolution*, 192.
85. Bergson, *Creative Evolution*, 141.
86. Bergson, *Creative Evolution*, 191.
87. In Bergson's words: "Now, since instinct is nowhere so developed as in the insect world, and in no group of insects so marvellously as in the Hymenoptera, it may be said that the whole evolution of the animal kingdom, apart from retrogressions towards vegetative life, has taken place on two divergent paths, one of which led to instinct and the other to intelligence" (*Creative Evolution*, 141).

88. Bergson, *Creative Evolution*, 192.
89. Bergson, *Creative Evolution*, 186.
90. Bergson, *Creative Evolution*, 174.
91. Bergson, *Creative Evolution*, 174, italics added.
92. Bergson, *Creative Evolution*, 186, italics added.
93. Bergson, *Creative Evolution*, 174.
94. Bergson, *Creative Evolution*, 186.
95. Whitehead, *Process and Reality*, 23.
96. Claire Petitmengin, "Enaction as a Lived Experience: Towards a Radical Neurophenomenology," *Constructivist Foundations* 12, no. 2 (2017): 143–45.
97. Bergson, *Creative Evolution*, 186.
98. Bergson, *Creative Evolution*, 174.
99. Bergson, *Creative Evolution*, 186, italics added.
100. Bergson, *Creative Evolution*, 187–88.
101. Bergson, *Creative Evolution*, 188.
102. Bergson, *Creative Evolution*, 195.
103. Ward Goodenough, "Navigation in the Western Carolines: A Traditional Science," in *Naked Science: Anthropological Inquiry into Boundaries, Power, and Knowledge,* ed. Laura Nader, 29–42 (New York: Routledge, 1996), 37–38.
104. Linda Groff, chap. 6 in this volume.
105. Theo Badashi, chap. 9 on this volume.
106. Jason James Kelly, chap. 8 in this volume.
107. Jason James Kelly, chap. 8 in this volume.
108. Bergson, *Creative Evolution*, 41–53, 133.
109. Theo Badashi, chap. 9 on this volume.
110. Theo Badashi, chap. 9 on this volume.
111. As Segall aptly says, "[o]nly very recently in the history of our species have these ritualized symbolic enactments become detached from their encompassing cosmic and biotic rhythms. Our modern myths have become too anthropocentric. We have immersed ourselves in a symbolic system that is radically out of tune with our ecological context"

(chap 2 in this volume).

112. Emile Durkheim, *The Elementary Forms of the Religious Life*, trans. Joseph Swain (New York: The Free Press, 1965).

113. In *Creative Evolution* spatialization is considered a real process that connects God and the world. Matter is conceived of as the final outcome of a divine act of emanation: the heterogeneous duration of creative energy that emanates from God gradually loses its tension and becomes increasingly homogeneous and thus determinate. The physical universe emerges out of a universal process of spatialization of divine energy into inorganic matter. Due to its homogeneous order, inorganic matter is ruled by laws that can be described mathematically without any reference to divine creativity. In this sense, the whole of inorganic material reality may be characterized as profane. Accordingly, in Bergson's processual cosmology the sacred-profane dichotomy may be given a metaphysical meaning by being read in parallel with the distinction God-inorganic matter (or divine duration-homogeneous order). Though not an emanationist thinker, Whitehead also claims that matter is continuously generated by physical acts of spatialization that depend on God. Each actual occasion manifests itself in time and space: it spatializes itself in the literal sense of that word. In Whitehead's metaphysics, however, the causal interpenetration of God and temporal actual occasions is conceived of in a more subtle way than in Bergson's *Creative Evolution*. God not only creates the so-called 'initial aim' of each temporal process (Whitehead, *Process and Reality*, 244), God also participates in all processes that compose the world (*Process and Reality*, 31, 88), being thus *influenced* by them. Therefore, from a Whiteheadian perspective the sacred-profane dichotomy cannot be paralleled with the distinction God-matter. Hence, from a broadly process metaphysical perspective that attempts to balance between Whitehead's and Bergson's philosophies, it is better to interpret the sacred-profane-dichotomy epistemologically rather than metaphysically. In other words, it is advisable to apply it only to a human conception of reality.

114. María Valeria Berros, "The Constitution of the Republic of Ecuador: Pachamama Has Rights," Environment & Society Portal, Arcadia 11 (2015); see Rachel Carson Center for Environment and Society: http://www.environmentandsociety.org/node/7131.

115. Republic of Ecuador, *Constitution of 2008*, Chapter 7, Article 71, accessed February 24, 2018: https://www.constituteproject.org

/constitution/Ecuador_2008.pdf.
116. Republic of Ecuador, *Constitution of 2008*, Chapter 7, Article 72.

CHAPTER SIX: GROFF

1. Part II of this article is an updated, expanded version of Linda Groff, "Future Human Evolution and Views of the Future Human: Technological Perspectives and Challenges," in *World Future Review,* Second Special Issue on "Future Human Evolution," Vol. 7 no. 2–3 (Summer/Fall 2015).
2. Linda Groff, "Models of Change: A Foresight Tool to Aid Policy-Makers," in *World Affairs: The Journal of International Studies,* Third Special Futures Issue 15, no. 4 (2011): 4, 12–38.
3. Christian Schwagerl, *The Anthropocene: The Human Era and How It Shapes Our Planet* (Sante Fe, NM: Synergetic Press, 2014).
4. Elizabeth Kolbert, *The Sixth Extinction: An Unnatural History* (New York: Picador, Henry Holt, 2014).
5. Alvin Toffler, *The Third Wave* (New York: Bantam Books, 1980); Linda Groff, "Social and Political Evolution" in *Encyclopedia of the Future* (New York: Macmillan, 1996).
6. Ray Kurzweil, *The Age of Spiritual Machines: When Computers Exceed Human Intelligence* (New York: Penguin Press, 1999); *The Singularity Is Near: When Humans Transcend Biology* (New York: Viking Press, 2005, and Penguin Press, 2006); *How to Create a Mind: The Secret of Human Thought Revealed* (New York: Penguin Books, 2012); *The Singularity Is Nearer* (forthcoming); Bart Koslo, *Heaven in a Chip: Fuzzy Visions of Society and Science in the Digital Age* (New York: Three Rivers Press, 1999).
7. Ray Kurzweil, *The Age of Spiritual Machines*; *The Singularity Is Near*; *How to Create a Mind*; *The Singularity Is Nearer*; "The Ubiquity and Predictability of the Exponential Growth of Information Technology—Talk at the Singularity Summit 2009" (October 6, 2009), accessed February 23, 2015, https://vimeo.com/7322310; Koslo, *Heaven in a Chip.*
8. See n.7.
9. Personal communication with the author.
10. Gregory Stock, *Metaman: The Merging of Humans and Machines into*

a Global Superorganism (New York: Simon & Schuster, 1993); *Redesigning Humans: Choosing Our Genes, Changing Our Future* (New York: Houghton Mifflin, 2002); Ramez Naam, *More Than Human: Embracing the Process of Biological Enhancement* (New York: Broadway Books, [2005] 2010).

11. Revin Koemar, "First Brain Transplant," YouTube (November 16, 2008), accessed April 17, 2017, https://www.youtube.com/watch?v=8On7rktFZME.

12. "Cyborg" definition, *Dictionary.com*, accessed February 12, 2014, http://www.dictionary.com/.

13. This concept of cyborgs or bionic human beings was popularized in the United States in the 1970s—and thereby entered the public imagination that far back—by two very popular science fiction television programs at the time. One was *The Six Million Dollar Man*. The other was *The Bionic Woman*. Both characters worked for U.S. Intelligence services and had severe accidents with physical injuries, which prompted them to become "bionic" characters who were able to use their extended superhuman powers in the service of justice as envisioned at the time. Another example of a cyborg occurs in *The Terminator* science fiction film series and franchise.

14. See Brian Nelson, "7 real-life human cyborgs," *Mother Nature Network* (2015), accessed September 4, 2015, http://www.mnn.com/leaderboard/stories/7-real-life-human-cyborgs; and Andrew Handley, "10 New Technologies That Will Make You a Cyborg," Listverse.com (August 3, 2013), accessed September 2, 2015, http://listverse.com/2013/08/03/10-new-technologies-that-will-usher-in-an-era-of-cyborgs/ for cyborg examples.

15. "Robot" definition, *Dictionary.com*.

16. Many positive, as well as negative, images of robots have been portrayed in science fiction, raising issues that humanity will eventually have to deal with in future. More positive robot examples include Data in *Star Trek: Next Generation,* and R2-D2 and C-3PO in *Star Wars*. More negative robot or computer examples include Data's evil twin brother in *Star Trek: Next Generation*, and Hal the Computer in *2001: A Space Odyssey*, to name just a few.

17. EPSRC-Engineering and Physical Science Research Council, "Principles of Robotics" (2011), accessed April 15, 2015, https://www.epsrc.ac.uk/research/ourportfolio/themes/engineering/activities/principlesof

robotics/; the Arts and Humanities Research Council, AHRC.

18. EPSRC, "Principles of Robotics."
19. See Mark Strauss, "Patrick Stewart on His Craft, 21st Century Science and Robot Ethics," *Smithsonian Magazine* (May 2014), accessed August 25, 2015, http://www.smithsonianmag.com/arts-culture/patrick-stewart-his-craft-21st-century-science-and-robot-ethics-180951153/.
20. In one very interesting episode ("The Measure of a Man") of *Star Trek: The Next Generation,* Star Fleet Command decided they wanted to decommission Data—an advanced android robot and important member of the Starship Enterprise crew—to study him. Captain Picard had to defend, in court, Data's right to continue to exist (and not be decommissioned) because he had become a sentient, i.e., self-aware being. Capt. Picard—played by Patrick Stewart—won the case, and Data was not decommissioned. See *Star Trek: Next Generation,* season 7, episode 9, "The Measure of a Man," written by Melinda M. Snodgrass, CBS, February 13, 1989. (See a somewhat related issue discussed under the next Artificial Intelligence section, based on the film *Ex Machina.*)
21. "Artificial Intelligence," *Merriam-Webster Online Encyclopedia,* accessed April 11, 2015, https://www.merriam-webster.com/dictionary/artificial%20intelligence.
22. Vangie Beal, "AI-Artificial Intelligence," Webopedia, accessed April 11, 2015, http://www.webopedia.com/TERM/A/artificial_intelligence.html.
23. Beal, "AI-Artificial Intelligence."
24. Ray Kurzweil, "Don't Fear Artificial Intelligence," *Time* (December 19, 2014) accessed February 23, 2018, http://time.com/3641921/dont-fear-artificial-intelligence/.
25. Kurzweil, "Don't Fear Artificial Intelligence."
26. Bill Joy, "Why the Future Doesn't Need Us," *Wired* (January 4, 2000), accessed February 23, 2018, https://www.wired.com/2000/04/joy-2/.
27. Jacob Kastrenakes, "These are the projects Elon Musk is funding to prevent killer AI," *The Verge* (July 3, 2015), accessed August 20, 2015, http://www.theverge.com/2015/7/3/8889515/elon-musk-funding-37-ai-research-projects.
28. Olivia Solon, "Elon Musk Says Humans Must Become Cyborgs to Stay Relevant. Is He Right?" *The Guardian* (February 15, 2017), accessed

February 23, 2018, https://www.theguardian.com/technology/2017/feb/15/elon-musk-cyborgs-robots-artificial-intelligence-is-he-right.

29. Staff and agencies, "Elon Musk Wants to Connect Brains to Computers With New Company [Neuralink]," *The Guardian* (March 28, 2017), accessed February 23, 2018, https://www.theguardian.com/technology/2017/mar/28/elon-musk-merge-brains-computers-neuralink.

30. *The American Heritage Science Dictionary* (New York: Houghton Mifflin, 2002).

31. Bruce H. Lipton, *The Biology of Belief: Unleashing the Power of Consciousness, Matter and Miracles*, 10th anniversary ed. (London: Hay House, [2005] 2015).

32. Karola Stotz, "With 'Genes' like That, Who Needs an Environment? Postgenomics's Argument for the 'Ontogeny of Information'," *Philosophy of Science* 73 (2006): 905–17; Peter Beurton, "Was sind Gene heute?" *Theory in Biosciences* 117 (1998): 90–99; Denis Noble, "Genes and Causation," *Philosophical Transactions of Royal Society* 366 (2008): 3001–15.

33. U.S., NIH, National Human Genome Institute, "Cloning Fact Sheet" (June 11, 2015), accessed June 30, 2015, https://www.genome.gov/25020028/.

34. Physics Forums, "Can Human Clones Reproduce?" *Physics Forums* (May 21, 2014), accessed June 30, 2015, https://www.physicsforums.com/threads/can-human-clones-reproduce.744315/.

35. Michigan State University, "Laws and Public Policy about Cloning," MSU Libraries, accessed June 30, 2015, http://staff.lib.msu.edu/skendall/cloning/laws.

36. U.S. Food and Drug Administration (FDA), "Myths about Cloning," accessed June 30, 2015, http://www.fda.gov/AnimalVeterinary/SafetyHealth/AnimalCloning/ucm055512.htm#Myth12.

37. James Hughes, "Transhumanist Position on Human Germline Genetic Modification," in Kurzweil, Accelerating Intelligence Blog (March 22, 2015), accessed February 23, 2018, http://www.kurzweilai.net/transhumanist-position-on-human-germline-genetic-modification.

38. Antonio Regalado, "Scientists Call for a Summit on Gene-Edited Babies," *MIT Technology Review* (March 19, 2015), accessed February 23, 2018, https://www.technologyreview.com/s/536021/scientists-call

-for-a-summit-on-gene-edited-babies/.

39. Hawk's Perch Technical Writing, LLC and UnderstandingNano.com, accessed February 23, 2018: http://www.understandingnano.com.

40. Engineering and Medicine, National Academies of Sciences; Division on Engineering and Physical Sciences; National Materials and Manufacturing Board; and Committee on Triennial Review of the National Nanotechnology Initiative *Triennial Review of the National Nanotechnology Initiative,* National Academies Press, March 6, 2017.

41. Earl Boysen, "Graphene: It's Applications and Uses," Hawk's Perch Technical Writing, LLC and UnderstandingNano.com, accessed February 23, 2018, http://www.understandingnano.com/graphene-applications.html.

42. K. Eric Drexler, *Engines of Creation: The Coming Era of Nanotechnology* (New York: Anchor Books, [1986] 1990).

43. See K. Eric Drexler, *Engines of Creation; Nanosystems: Molecular Machinery, Manufacturing, and Computation* (New York: John Riley & Sons, 1992); *Radical Abundance: How a Revolution in Nanotechnology Will Change Civilization* (New York: Public Affairs, 2013); Eric K. Drexler, Chris Peterson, Gayle Pergamit, *Unbounding the Future: Nanotechnology Revolution* (New York: Simon & Schuster, 1992); J. Storrs Hall, *Nanofuture: What's Next For Nanotechnology* (Amherst, NY: Prometheus Books, 2005); Linda D. Williams, *Nanotechnology Demystified,* 1st edition (New York: McGraw-Hill, 2007).

44. Teena Hammond, "CISCO: The Internet of Everything is at Tipping Point," *Tech Republic* (February 18, 2015), accessed August 25, 2015, http://www.techrepublic.com/article/cisco-the-internet-of-everything-is-at-tipping-point/.

45. Hammond, "CISCO: The Internet of Everything is at Tipping Point."

46. Hammond, "CISCO: The Internet of Everything is at Tipping Point."

47. Luke Dormehl, "Google's Eric Schmidt Predicts the Disappearance of the Internet," *Fast Feed, c/o Fast Company* (January 23, 2015), accessed August 25, 2015, https://www.fastcompany.com/3041343/fast-feed/googles-eric-schmidt-predicts-the-disappearance-of-the-internet.

48. Dormehl, "Google's Eric Schmidt Predicts the Disappearance of the Internet."

49. Space X, www.spacex.com, accessed August 25, 2015.

50. Klint Finley, "Internet by Satellite is a Space Race with No Winners," *Wired* (June 12, 2015), accessed August 25, 2015, http://www.wired.com/2015/06/elon-musk-space-x-satellite-internet/; Chris Anderson, "Elon Musk's Mission to Mars," *Wired* (October 12, 2012). accessed August 25, 2015, https://www.wired.com/2012/10/ff-elon-musk-qa/.

51. c/net. *Virtual Reality 101,* accessed February 23, 2018, https://www.cnet.com/special-reports/vr101/.c/net.

52. c/net, *Virtual Reality 101;* Murray Ramirez, *Virtual Reality for Beginners: How To Understand, Use & Create With VR* (North Charleston, SC: CreateSpace Independent Publishing Platform, 2016).

53. Ramirez, *Virtual Reality for Beginners.*

54. Steve Aukstakalnis, *Practical Augmented Reality: A Guide to the Technologies, Applications, and Human Factors for AR and VR (Usability)* (Boston: Pearson Education, 2017).

55. Christopher Barnatt, *3D Printing* 3rd ed. (CreateSpace Independent Publishing Platform, 2016); Chee Kai Chua, Kah Fai Leong, *3D Printing and Additive Manufacturing: Principles and Applications: 5th Edition of Rapid Prototyping* (Singapore: World Scientific Publishing, 2017); Ian Gibson, David Rosen, Brent Stucker, *Additive Manufacturing Technologies: 3D Printing, Rapid Prototyping, and Direct Digital Manufacturing,* 2nd edition (New York: Springer, 2015); Raymond T. Reeves, *Additive Manufacturing Technologies: 3D Printing, Rapid Prototyping, and Direct Digital Manufacturing* (LM Publishing, Amsterdam, 2015).

56. Barnatt, *3D Printing;* Chua, Leong, *3D Printing and Additive Manufacturing*; Gibson, Rosen, Stucker, *Additive Manufacturing Technologies;* Reeves, *Additive Manufacturing Technologies.*

57. NASA, "NASA's Journey to Mars" (December 1, 2014), accessed August 25, 2015, https://www.nasa.gov/content/nasas-journey-to-mars.

58. SpaceX, www.spacex.com; Anderson, "Elon Musk's Mission to Mars."

59. Stichting Mars One, Mars One Ventures PLC, *Mars One—Human Settlement on Mars,* accessed August 25, 2015: http://www.mars-one.com.

60. Graham Templeton, "Mars One Candidates to Compete for Place on Imaginary Mission," *Extreme Tech* (July 20, 2015), accessed August 25, 2015, http://www.extremetech.com/extreme/210270-mars-one-candidates-to-compete-for-place-on-imaginary-mission.

61. See my forthcoming book, *Options for Future Human Evolution,* for

a chapter on the many crises currently facing Earth as we enter the Anthropocene Age characterized by human dominance of the Earth.

62. Alfred North Whitehead, *Process and Reality*, corrected edition, ed. David Ray Griffin and Donald W. Sherburne (New York: The Free Press, 1978); Ivor Leclerc, *Whitehead's Metaphysics* (Bloomington: Indiana University Press, 1975); Victor Lowe, *Understanding Whitehead* (Baltimore. MD: John Hopkins Press, 1966); Donald W. Sherburne, *A Whiteheadian Aesthetic* (New Haven, CT: Yale University Press, 1961); John B. Cobb, Jr., *Whitehead Word Book* (Claremont, CA: P&F Press, 2008); William A. Christian, *An Interpretation of Whitehead's Metaphysics* (New Haven, CT: Yale University Press, 1967); Dorothy Emmet, *Whitehead's Philosophy of Organism* (Westport, CT: Greenwood Press, 1981); John W. Lango, *Whitehead's Ontology* (Albany: State University of New York Press, 1972).

CHAPTER SEVEN: MACCRACKEN

1. The author thanks the editors of this volume for their enthusiasm for this adventurous and peculiar study. Thanks go to Matthew Segall for providing important connections when they were needed. Gratitude to my advisor at the University of Virginia, Sonam Kachru, whose extensive feedback on early preparatory work for this study improved the final result considerably. Thanks also to Neha Chriss, for her many inspiring conversations and valuable insights from a STEM perspective, and for directing my attention toward the work of Michael Graziano. All imperfections naturally remain the author's own. May this work, though a small endeavor, spark the fire of transdisciplinary conversations and cooperation, and may such interactions be of benefit to all beings!

2. It must be noted that a broad definition is here intended, in line with the Oxford English Dictionary's (OED) entry for transhumanism: "A belief that the human race can evolve beyond its current limitations, esp. by the use of science and technology." Note, however, that such broad usage may result in an etic designation of certain thinkers (e.g., Kurzweil) as transhumanist—due to the overwhelming currency of their ideas in transhumanist discourse—whether or not such a person would emically self-apply the term "transhumanist" unreservedly. Note also that the OED's definition bears the direct influence of transhumanist Dr. Max More, who was consulted for the entry.

3. Hank Pellisier and Teresa dal Santo, "Transhumanists: Who Are They?" *Journal of Personal Cyberconsciousness* 7, no. 2 (2012): 20–29, accessed October 28, 2016: http://www.terasemjournals.org/PCJournal/PC0801/Papers/Pellissier_Dal_Santo_APA.pdf. Complete survey data is also available from the authors upon request.

4. For the Technoprogressive Declaration, see http://ieet.org/index.php/IEET/more/tpdec2014 (accessed March 2017); See also: James Hughes, "Technoprogressive Biopolitics," in *Progress in Bioethics*, ed. Jonathon Moreno and Sam Berger (Cambridge: MIT Press, 2009).

5. Buddhist Epistemology here specifically denotes the textual tradition of Vasubandhu, Dignāga, and Dharmakīrti, all of whom are associated with Vijñanavāda idealism, but who also set in place a number of guidelines concerning knowledge and debate that have exerted considerable influence on Buddhist thought more generally. For an introduction strongly grounded in primary sources, see William Edelglass and Jay L. Garfield, eds., *Buddhist Philosophy: Essential Readings* (New York: Oxford University Press, 2009), 261–369.

6. Concerning the threads of rationality within Buddhist discourse and its complex relationship with science-based worldviews, see Donald S. Lopez, *Buddhism and Science: A Guide for the Perplexed* (Chicago: University of Chicago Press, 2008).

7. One standout example is to be found in the Global Future 2045 Conference series, which has made an effort to engage the ideas of both the indigenous practice community, as represented by the Dalai Lama, and Tibetanist Robert Thurman. See GF2045, http://gf2045.com/.

8. There is an advised distinction here between absolute and epistemic idealism, but one that represents a needless digression for present purposes. Suffice to say, contemporary scholarship generally agrees that the epistemologist Dharmakīrti deployed what Dunne, *Foundations of Dharmakīrti's Philosophy* has called a "sliding scale of analysis." However, scholars are by no means in agreement concerning what, for Dharmakīrti, was the nature of the subtlest strata of metaphysical outlook, for which see Dan Arnold, "Buddhist Idealism, Epistemic and Otherwise: Thoughts on the Alternating Perspectives of Dharmakīrti," *Sophia* 47 (2008): 3–28, accessed October 22, 2016, DOI: 10.1007/s11841-008-0046-7; also Birgit Kellner, "Dharmakīrti's Criticism of External Realism and the Sliding Scale of Analysis," in *Religion and Logic in Buddhist Philosophical Analysis. Proceedings of the Fourth International Dharmakīrti Conference, Vienna: August 23–27, 2005,*

9. Glosses here are adopted from Cabezón, "Language and the Ultimate: Do Mādhyamikas Make Philosophical Claims: A Selection from Khedrupjey's Great Digest (Stong thun chen mo)." In: William Edelglass and Jay Garfield, eds., *Buddhist Philosophy: Essential Readings* (Oxford: Oxford University Press, 2009), 126–37. The phrase "textual tradition" is worth commenting on, for those outside philological disciplines. Such a phrase is used as opposed to "school," only to emphasize a peculiar tensile strength within premodern traditions of commentary: that tension between appeal to the weight of traditional authority vs. the impulse to modify and innovate in pursuit of a more trenchant and subtle articulation of philosophical truth. In brief, "textual tradition" is meant to emphasize a diverse, evolving intellectual environment and lineage, vs. the possible dogmatic connotation of "school."

10. Such an assertion is a key thesis of Herbert V. Guenther, *From Reductionism to Creativity* (Boston: Shambhala, 1989).

11. "Dr. Robert Thurman—Inner Science for Merging our Cybernetic and Subtle Bodies," accessed March, 2017, https://www.youtube.com/watch?v=6YMokHYF8cI&.

12. Thurman, "Inner Science." Thurman's above discussion entails a critique of what I would term a scientistic dogma of refuting conclusions based on absence of evidence. This stands in contrast to any philosophical discourse that seeks to distinguish between what can and cannot be known to what degree of certainty. Indeed, such scientistic dogma offends principled adherence to scientific method itself, by using incomplete evidence in support of a definitive conclusion.

13. Max More, "The Philosophy of Transhumanism," in *The Transhumanist Reader: Classical and Contemporary Essays on the Science, Technology, and Philosophy of the Human Future*, ed. Max More and Natasha Vita-More (Chichester: Wiley-Blackwell, 2013), 7.

14. Derek Parfit, *Reasons and Persons* (Oxford: Clarendon Press, 1984), x.

15. More, "The Diachronic Self: Identity, Continuity, Transformation," Doctoral Dissertation (University of Southern California, 1995), 159–61. Though offering critiques of More's ideas here, the author wishes to extend gratitude to Dr. More for sharing his dissertation during a time when it was not readily accessible. Cited page numbers correspond to the now publicly available edition.

16. Parfit, *Reasons and Persons*, 206.
17. For morphological freedom, see More, "The Philosophy of Tranhumanism," 4, 15.
18. More, "The Diachronic Self," 159–64.
19. For the generalized concept of "mind uploading"—to other bodies, machines, or other "substrates," see Randal A. Koene, "Substrate-Independent Minds"; Giulio Prisco, "Transdendent Engineering"; and Martine Rothblatt, "Deeper than Matter," all three found in *The Transhumanist Reader: Classical and Contemporary Essays on the Science, Technology, and Philosophy of the Human Future*, ed. Max More and Natasha Vita-More (Chichester, UK: Wiley-Blackwell, 2013).
20. Parfit, *Reasons and Persons*, 199–347.
21. See also Nigel Tetley, "The Doctrine of Rebirth in Theravāda Buddhism: Arguments for and Against," Doctoral Dissertation (University of Bristol, 1990) 288, 295.
22. Parfit, *Reasons and Persons*, 207–08.
23. Parfit, *Reasons and Persons*, 207–08.
24. Parfit, *Reasons and Persons*, 208–09.
25. The name Omega Point Reconstruction may well be Dr. More's sly nod in the direction of Teilhard de Chardin, whose teleological Omega Point looms in the background of transhumanist metaphysics. See James Hughes, *Citizen Cyborg* (Cambridge MA: Westview Press, 2004), 174; and David Brin et al., "Vinge's Singularity Concept," in *The Transhumanist Reader: Classical and Contemporary Essays on the Science, Technology, and Philosophy of the Human Future*, ed. Max More and Natasha Vita-More (Chichester, UK: Wiley-Blackwell, 2013), 396.
26. More, "The Diachronic Self," 15–16.
27. Parfit, *Reasons and Persons*, 255–56.
28. Note that the idea of branching introduces the notion of the Closest-Continuer (CC) view. For example, in the TT scenario, if the transporter produces a duplicate without disintegrating the original, the original is the CC, until the event of his or her death makes the duplicate the CC—the inheritor of the original, pre-transporter personhood of the original person. The designation *patternist* is indebted to Loriliai Biernacki, "A Cognitive Science View of Abhinavagupta's Understanding of Consciousness," *Religions* 5 (2014): 767–79, accessed

October 28, 2016, DOI: 10.3390/rel5030767, which highlights a comparison between Buddhist and materialist thinkers, stating "Buddhist models of *anātman*, 'no-self' doctrine sound a lot more like patternist conceptions, like Daniel Dennett's or Ray Kurzweil's . . . or in some cases, informationalist conceptions."

29. More, "The Diachronic Self," 12.
30. More, "The Diachronic Self," 23.
31. Such a sweeping statement necessitates some explanation, if not defense. The phrase "artificial human bodies" is here meant to denote the sort of android body that would effectively mimic the average human body and could conceivably pass a Turing test. This is not to be confused with the sort of human-skinned "service" bots being developed primarily in Japan by the likes of Hiroshi Ishihuro Laboratories, which at the time of this writing are still contending with the "uncanny valley" effect (http://www.geminoid.jp, accessed March, 2017). Brain transplants, likewise, while they may be technically possible, run into a number of serious logistical and ethical obstacles in becoming conventionally possible. There is no current project, to the author's knowledge, concerned with developing the "decerebrated clone" that More hypothesizes. Relative to the above scenarios, a personhood controversy arising from the advent of progressively more sophisticated AI systems—likely not embodied ones—seems the more likely scenario. Note that at the time of this writing, rudimentary artificial life is also possible. It may well be that the convergence of the AI, nanotech, and biotech revolutions does indeed precipitate a technological phase transition not unlike the Singularity of transhumanist speculative metaphysics, as in Kurzweil, *The Singularity is Near: When Humans Transcend Biology* (New York: Viking, 2005); and *The Age of Spiritual Machines: When Computers Exceed Human Intelligence* (New York: Viking, 1999). In such a case, brain transplants, artificial bodies, human cloning, or partial brain replacements may outpace the importance of strong AI based on brain scanning. But based on current trends, the latter appears the most likely to be implicated in personhood controversy.
32. In the United States, at least, some limited precedent for this controversy exists—and relates the philosophy under consideration to the Technoprogressive Declaration—in the legal controversy over so-called "corporate personhood." That is, a sharp divide exists between Libertarian and Democratic Socialist agendas with respect to human rights such as free speech (under the First

Amendment) granted not only to private individuals but to corporations. An entirely distinct consideration concerning AIs with respect to legal practice is the fiduciary responsibility of a law office to its client. As one anonymous colleague in the study of law has pointed out, as strong AI develops, it may become increasingly difficult to justify to clients the potential detriment that might come about as a result of *not* deploying the advanced computing power of AI in the preparation of a case.

33. More, "The Diachronic Self," 210.
34. Parfit, *Reasons and Persons*, 347.
35. That is, More was a consultant on the *Oxford English Dictionary's* entry, "Tranhumanism," which should help explain why the reader of the OED is referred by that very entry to Max More's brand "Extropianism."
36. More, "The Philosophy of Transhumanism," 7.
37. More, "The Philosophy of Transhumanism," 8.
38. More, "The Philosophy of Transhumanism," 7.
39. Philosophers of process might wish to critique Parfit's *Relation R* in terms of its occasional reduction of the qualitative to the quantitative (e.g., in imagining numerical values for traits like creativity—something More expands upon). To take but one influential twentieth century example, Giles Deleuze's *Le Bergsonisme* (Paris: Presses universitaires de France, 1966) is, in part, dedicated to describing Bergson's method of *intuition*, wherein Bergson is understood to privilege the qualitative over the quantitative, with the latter being merely an extraction of the former. There is, in the method of *intuition* in Deleuze's Bergson, a metaphysical inversion of the objectifying quality perceived in a preoccupation with the quantitative. In such a metaphysics, it would be dangerously misguided to attempt to assign numerical values to subjective and qualitative attributes such as creativity and so forth.

    However, such critique is intentionally bracketed here. As stated previously, it is the intention of this chapter to think along with Parfit and to discover common vocabulary, rather than attempt a deconstruction of his assumptions. Moreover, to pursue a deconstruction in this case does not successfully undermine Parfit's philosophy. For it is precisely Parfit's point that the numerical values are to be variously interpreted. To argue that the numbers themselves that were assigned are in the first place arbitrary is an undertaking that perhaps only reinforces Parfit's point, or perhaps obscures the issue at hand.

40. Parfit, *Reasons and Persons*, 273.
41. Parfit, *Reasons and Persons*, 280.
42. Steven Collins, *Selfless Persons: Imagery and Thought in Theravāda Buddhism* (New York: Cambridge University Press, 1982), 177.
43. Tetley, "The Doctrine of Rebirth," 228-296, esp. 231-232.
44. Steven Collins, "A Buddhist Debate about the Self; And Remarks on Buddhism in the Work of Derek Parfit and Galen Strawson" *Journal of Indian Philosophy* 25 (1997): 467–93, accessed October 22, 2016, DOI: 10.1023/A:1004287006506.
45. Paul Williams, *Altruism and Reality: Studies in the Philosophy of the Bodhicaryāvatāra* (Surrey: Curzon Press UK, 1998). Note also that "Mādhyamika" is the adjectival designation for the noun, "Madhyamaka."
46. One notable exception is Jim B. Tucker, *Return to Life: Studies in the Philosophy of the Bodhicaryāvatāra* (Surrey, UK: Curzon Press, 1998) and *Life Before Life: Studies in the Philosophy of the Bodhicaryāvatāra* (Surrey, UK: Curzon Press, 1998). Tucker's work carries on the work of Bruce Greyson, preceded by Ian Stevenson at the University of Virginia. Their Division of Perceptual Studies unit is perhaps the preeminent example of a major research university concerned with the study of past life memory phenomena. Naturally, their research awaits peer review and corroboration, let alone acceptance within mainstream medical science (accessed March, 2017,) https://med.virginia.edu/perceptual-studies/who-we-are/.
47. Mark Siderits, "Buddhist Reductionism," *Philosophy East and West* 47, no. 4 (1997): 455–78, accessed October 28, 2016, DOI: 10.2307/1400298.
48. Jim Stone, "Parfit and the Buddha: Why There Are No People," *Philosophy and Phenomenological Research* 48, no. 3 (1988): 519–32, accessed October 28, 2016, http://www.jstor.org/stable/2107477, and James Giles, "The No-Self Theory: Hume, Buddhism, and Personal Identity," *Philosophy East and West* 43 (1993): 175–20, accessed October 22, 2016, http://www.jstor.org/stable/1399612.
49. Siderits, "Buddhist Reductionism," 249.
50. Derek Parfit, "The Unimportance of Identity," in *Identity: Essays Based on Herbert Spencer Lectures Given in the University of Oxford* (New York: Oxford University Press, 1995), 16.
51. Parfit, "The Unimportance of Identity," 17.

52. Siderits, "Buddhist Reductionism," 232.
53. Siderits, "Buddhist Reductionism," 235.
54. Jonardon Ganeri, *The Concealed Art of the Soul: Theories of Self and Practices of Truth in Indian Ethics and Epistemology* (New York: Oxford University Press, 2007), 160–66.
55. Ganeri, *The Concealed Art of the Soul*, 162, The passage in question appears in Parfit, *Reasons and Persons*, 502 and Derek Parfit, "Experiences, Subjects, and Conceptual Schemes," *Philosophical Topics* 26:1/2 (1999): 217–70, 260, accessed October 28, 2016, http://www.jstor.org/stable/43154286.
56. Ganeri, *The Concealed Art of the Soul*, 163.
57. Ganeri, *The Concealed Art of the Soul*, 163.
58. James Duerlinger, *Indian Buddhist Theories of Persons: Vasubandhu's "Refutation of the Theory of a Self"* (New York: RoutledgeCurzon, 2003), 73-74.
59. Ganeri, *The Concealed Art of the Soul*, 165–66.
60. Nāgārjuna, *The Fundamental Wisdom of the Middle Way: Nāgārjuna's Mūlamadhyamakakārikā*, trans. Jay L. Garfield (New York: Oxford University Press, 1995), 3, 103–07.
61. Roy W. Perrett, "Personal Identity, Minimalism, and Madhyamaka," *Philosophy East and West* 52, no.3 (2002): 373–85.
62. Mark Johnston, "Reasons and Reductionism," *The Philosophical Review* 101, no. 3 (1992): 589–618, http://www.jstor.org/stable/2186058.
63. Michhael S.A. Graziano, *Consciousness and the Social Brain* (New York: Oxford University Press, 2013), 47.
64. Graziano, *Consciousness*, 202–08.

CHAPTER EIGHT: KELLY

1. Lynn White, Jr.,"The Historical Roots of the Ecological Crisis," *Science* 155 (March 10, 1967).
2. Kenneth Brower, "The Dangers of Cosmic Genius," *The Atlantic*, October, 27, 2010.
3. James Gustave Speth, *America the Possible: Manifesto for a New Economy* (New Haven, CT: Yale University Press, 2013), 194.
4. See Leslie Sponsel, *Spiritual Ecology: A Quiet Revolution* (Santa Barbara,

CA: Praeger, 2012).

5. See David Landis Barnhill and Roger S. Gottlieb, *Deep Ecology and World Religions: New Essays on Sacred Ground* (Albany: State University of New York Press, 2001).
6. Loyal Rue, *Religion is Not about God* (New Brunswick, NJ: Rutgers University Press, 2005), 25.
7. Edward Carpenter, *From Adam's Peak to Elephanta* (London: Swan Sonnenschein, 1892), 154.
8. Carpenter, *From Adam's Peak to Elephanta*, 176.
9. Carpenter, *From Adam's Peak to Elephanta*, 176.
10. For a more comprehensive discussion of these issues see Sheila Rowbotham, *Edward Carpenter: A Life of Liberty and Love* (New York: Verso, 2009).
11. Edward Carpenter, *My Days and Dreams* (London: George Allen & Unwin, 1921), 190.
12. Carpenter, *My Days and Dreams*, 106.
13. Walt Whitman as quoted by David S. Reynolds, *Walt Whitman* (London: Oxford University Press, 2005), 96.
14. Whitman, *Leaves of Grass* (New York: Penguin Books, [1855] 1986), 44.
15. For a comprehensive discussion of the controversy surrounding Whitman and homoeroticism, see Gary Schmidgall, *Walt Whitman: A Gay Life* (New York: Plume, 1998).
16. Edward Carpenter, *Days with Walt Whitman* (London: George Allen & Unwin, 1906), 56.
17. Edward Carpenter, *Towards Democracy* (Charleston: Nabu Press, [1883] 2011), 1.
18. Carpenter, *Towards Democracy*, 111.
19. Carpenter, *Towards Democracy*, 11.
20. Carpenter, *Towards Democracy*, 21.
21. Carpenter, *Towards Democracy*, 41.
22. Carpenter, *Towards Democracy*, 6.
23. Edward Carpenter, *Civilization: Its Cause and Cure* (London: Humboldt Publishing, 1891), 39.

24. Edward Carpenter, *The Art of Creation* (London: George Allen & Unwin, 1904), 54.
25. Carpenter, *The Art of Creation*, 59.
26. Carpenter, *The Art of Creation*, 60.
27. Carpenter, *Civilization: Its Cause and Cure*, 52.
28. William James, *The Varieties of Religious Experience* (New York: Penguin Books, [1901] 1958), 320.
29. James, *The Varieties of Religious Experience*, 341.
30. R.M. Bucke, *Cosmic Consciousness* (New Jersey: Citadel Press, [1901] 1973), 8.
31. For a more comprehensive discussion of Bucke and Carpenter's relationship, see Lorna Weir's "Cosmic Consciousness and the Love of Comrades: Contacts Between R.M. Bucke and Edward Carpenter," *Journal of Canadian Studies* 30.2 (1995).
32. Bucke, *Cosmic Consciousness*, 2.
33. Bucke, *Cosmic Consciousness*, 14.
34. Bucke, *Cosmic Consciousness*, 58.
35. Bucke, *Cosmic Consciousness*, 61.
36. Bucke, *Cosmic Consciousness*, 63.
37. Jeffrey J. Kripal, *Esalen: America and the Religion of No Religion* (Chicago: University of Chicago Press, 2007), 419.
38. This idea of "evolutionary mysticism" fits well with what Linda Groff identifies as "Evolving Consciousness-Based Perspectives." See Groff's "Future Human Evolution and Views of the Future Human: Technology, Sustainability, & Consciousness-Based Perspectives," in this volume.
39. Bucke, *Cosmic Consciousness*, 123.
40. Whitman, *Leaves of Grass*, 25.
41. This ethical position accords well with the views of deep/spiritual ecology. Consider, for example, the following observation by Arne Naess:

> the greater our comprehension of togetherness with other beings, the greater the identification and the greater care we will take. The road is also opened thereby for delight in the wellbeing of

others and sorrow when harm befalls them. We see what is best for ourselves, but through the extension of self, our 'own' best is also that of others. The own/not own distinction survived only in grammar, not in feeling (Arne Naess, *Ecology, Community and Lifestyle*, trans. David Rothenberg [Cambridge: Cambridge University Press, 1989], 175).

42. Richard Sylvan, "A Critique of Deep Ecology," *Radical Philosophy* 40, no. 2–12 (1985).

43. Clare Palmer, *Environmental Ethics and Process Thinking* (London: Clarendon Press, 1998), 173.

44. Alfred North Whitehead, *Modes of Thought* (New York: Macmillan, 1938), 29.

45. Palmer, *Environmental Ethics and Process Thinking*, 201.

46. John B. Cobb Jr. and David Ray Griffin, *Process Theology: An Introductory Exposition* (Belfast: Christian Journals, 1977), 117.

47. Alfred North Whitehead, *Adventures of Ideas* (Harmondsworth. UK: Pelican, 1933), 226.

48. Whitehead, *Modes of Thought*, 191.

49. John B. Cobb Jr., "What God Does," in *Back to Darwin: A Richer Account of Evolution*, ed. John Cobb B. Jr. (Grand Rapids, MI: William B. Eerdmans Publishing Company, 2008), 410.

50. Cobb, "What God Does," 410.

51. Matthew T. Segall, "Religion in Human and Cosmic Evolution: Whitehead's Alternative Vision," in this volume.

52. Blair R. Reynolds, "Cosmic Ecstasy and Process Philosophy," *The Journal of Natural and Social Philosophy* 1, no. 2 (2005): 332.

53. Reynolds, "Cosmic Ecstasy and Process Philosophy," 332.

54. Reynolds, "Cosmic Ecstasy and Process Philosophy," 331.

55. I believe a parallel can be drawn here between my conceptualization of cosmic consciousness and Spyridon A. Koutroufinis' idea of *"ecological intuition"* as outlined in his essay from this volume.

56. Alfred North Whitehead, *Process and Reality: An Essay in Cosmology*, corrected ed,. ed. David Ray Griffin and Donald W. Sherburne (New York: Free Press, 1978), 351.

CHAPTER NINE: BADASHI

1. Though I have clearly drawn upon many sources for much of my own philosophical views, a significant number of the propositions I put forth came through my own phenomenological and holotropic explorations, and when I say that I cannot take credit for any of them myself, as I am a mode of the earth, I truly mean that. However, it cannot be emphasized enough how deeply my own work has been impacted by the following teachers, allies and friends: Brian Swimme, Richard Tarnas, Robert McDermott, Joanna Macy, Sean Kelly, Thomas Berry, Steve Mc Intosh, Catherine Keller, Charles Eisenstein, Kevin Kelly, Mathew Tarnas Segall, and my perpetual co-conspirator Maximilian DeArmon.

2. W. Brian Arthur, *The Nature of Technology: What It Is and How It Evolves* (New York: Free Press, 2009), 22.

3. Edmund J. Bourne, *Global Shift: How a New Worldview Is Transforming Humanity* (Petaluma, CA: Noetic Books, Institute of Noetic Sciences, 2008).

4. I am not necessarily arguing that these findings suggest the existence of an omnipotent Creator God consciously directing the evolution of the universe, only that the ubiquitous and mathematically improbable conditions that permit life to exist do indeed suggest the presence of what we could call Cosmic Intelligence.

5. Richard Tarnas, *"Cosmos and Psyche,"* Seminar, California Institute of Integral Studies, Spring, 2015.

6. Alfred North Whitehead, *Science and the Modern World* (Cambridge, UK: Cambridge University Press, [1925] 1960), 178.

7. Harari, Yuval N. *Sapiens: A Brief History of Humankind* (New York: Harper, 2015).

8. For a detailed look at the philosophy of cosmic consciousness, and how it intersects with Whitehead's philosophy of organism, please read Jason James Kelly's wonderful chapter *Shared Vulnerabilities: Cosmic Consciousness and the Philosophy of Organism*, in this volume.

For Further Reading

Benyus, Janine M. *Biomimicry: Innovation Inspired by Nature.* New York: Morrow, 1997.

Berry, Thomas. *The Dream of the Earth.* San Francisco: Sierra Club Books, 1988.

Berry, Thomas. *The Great Work: Our Way into the Future.* New York: Bell Tower, 1999.

Eisenstein, Charles. *The Ascent of Humanity: Civilization and the Human Sense of Self.* Berkeley, CA: Evolver Editions, 2013.

Gardner, James N. *Biocosm: The New Scientific Theory of Evolution : Intelligent Life Is the Architect of the Universe.* Makawao, Maui, HI: Inner Ocean, 2003.

Harari, Yuval N. *Sapiens: A Brief History of Humankind.* New York: Harper, 2015.

Hargens, Sean, Michael E. Zimmermann. *Integral Ecology: Uniting Multiple Perspectives on the Natural World.* Boston: Integral Books, 2009.

Keller, Catherine. *From a Broken Web: Separation, Sexism, and Self.* Boston: Beacon, 1986.

Kelly, Kevin. *What Technology Wants.* New York: Viking, 2010.

Laszlo, Ervin. *Thomas Berry, Dreamer of the Earth: The Spiritual Ecology of the Father of Environmentalism.* Rochester, VT: Inner Traditions, 2011.

McDonough, William, and Michael Braungart. *The Upcycle: Beyond Sustainability—Designing for Abundance.* New York: North Point Press, 2013.

McIntosh, Steve. *Integral Consciousness and the Future of Evolution: How the Integral Worldview Is Transforming Politics, Culture, and Spirituality.* St. Paul, MN.: Paragon House, 2007.

Mickey, Sam. *On the Verge of a Planetary Civilization A Philosophy of Integral Ecology.* London: Rowman & Littlefield International, 2014.

Swimme, Brian. *The Universe Is a Green Dragon: A Cosmic Creation Story.* Santa Fe, NM: Bear, 1985.

Swimme, Brian, Thomas Berry. *The Universe Story: From the Primordial Flaring Forth to the Ecozoic Era—A Celebration of the Unfolding of the Cosmos.* San Francisco: Harper SanFrancisco, 1994.

Tarnas, Richard. *The Passion of the Western Mind: Understanding the Ideas That Have Shaped Our World View.* New York: Ballantine Books, 1993.

CONTRIBUTORS

Theodore Badashi, M. A.
Santa Fe, NM 87501
USA

Terrence W. Deacon, Ph. D., Professor
Anthropology Department, University of California, *Berkeley*
Berkeley, CA 94720
USA

Alex Gomez-Marin, Ph. D.
Behavior of Organisms Laboratory
Instituto de Neurociencias CSIC-UMH
Sant Joan d'Alacant,
03550 Alicante
Spain

Linda Groff, Ph. D., Professor Emeritus
Director, Global Options & Evolutionary Futures Consulting
California State University, Dominguez Hills
Carson, CA 90747
USA

Jason James Kelly, Ph. D., Professor
School of Religion, Theological Hall
Queen's University
K7L 3N6 Kingston, ON
Canada

Spyridon A. Koutroufinis, Dr., Privatdozent
Institute of Philosophy, History of Literature, Science, and Technology
Technische Universität Berlin
10623 Berlin
Germany

Sean MacCracken, M. A.
Asian Philosophies and Cultures
California Institute of Integral Studies
San Francisco, CA 94103
USA

René Pikarski, M. A.
Munich School of Philosophy
80539 Munich
Germany

Matthew T. Segall, Ph. D.
Philosophy, Cosmology, and Consciousness Program
California Institute of Integral Studies
San Francisco, CA 94103
USA

Index of Persons

Adorno, Theodor W., 47
Alberch, Pere, 65
Amkreutz, Jan, 128, 144, 145
Araujo, Arthur, 96
Aristarchus of Samos, 214
Aristotle, 130, 226
Asimov, Isaac 135, 136

Bacon, Francis, 215
Badashi, Theo 10, 106, 118, 119, 263
Bellah, Robert, 6, 19-22, 24-32
Benyus, Janine, 210, 235
Bergson, Henri, 6-8, 37, 38, 41-50, 53, 56, 63, 66, 68-70, 72-75, 79-83, 86-89, 91, 98, 100-107, 109-114, 116-119, 247, 248, 250, 251, 253, 264, 276
Berry, Thomas, 207, 210, 223, 232, 236, 237, 241, 282
Bezos, Jeffrey, 151
Blake, William, 34
Braungart, Michael, 239
Buber, Martin, 29, 31, 32

Bucke, Maurice, 9, 118, 187, 193-198, 200, 202, 237
Buddha (Siddhārtha Gautama), 158, 159, 173-176, 181, 182, 196

Campbell, Joseph, 121
Čapek, Milič, 248
Carpenter, Edward, 9, 118, 185, 187-192, 195-198, 200, 201, 237
Cassirer, Ernst, 8, 37, 45, 91, 98, 99, 101-103, 106, 107, 109, 111
Clarke, Arthur C., 133
Cobb, John B., Jr., 199, 201
Copernicus, Nicolaus, 214, 215

Dalai Lama, 272
d'Alembert, Jean le Rond, 36
Darwin, Charles, 20, 24, 64, 88, 90, 100, 140, 141, 151, 215
Dawkins, Richard, 20, 21, 31, 215
De Chardin, Pierre Teilhard, 145, 197, 219, 232, 274
Deacon, Terrence William, 6, 24, 45, 131

Index of Persons

Deleuze, Gilles, 49, 276
Demetrius the Cynic, 57
Dennett, Daniel, 21, 22, 24, 26, 31, 32, 275
Descartes, René, 33, 215, 257
Drexler, Eric 143, 144
Durkheim, Emile, 119

Emerson, Ralph Waldo, 189

Feyerabend, Paul, 85
Fichte, Johann Gottlieb, 197
Foucault, Michel, 7, 37, 38, 43, 49-55, 57, 58
Freud, Sigmund, 135

Galilei, Galileo, 215
Gates, Bill 139, 148
Geertz, Clifford, 29, 31
Gomez-Marin, Alex, 7, 42, 106, 116, 246, 247
Griffin, David Ray, 199
Groff, Linda, 8, 61, 117

Hegel, Georg Wilhelm Friedrich, 4, 197, 237
Heraclitus of Ephesus, 82
Horkheimer, Max, 47
Huizinga, Johan, 26, 27, 29, 34
Husserl, Edmund, 47, 50, 108, 109, 111

Ingold, Tim, 66
James, William, 22, 118, 193, 194, 196, 200
Jesus, 196
Jonas, Hans, 111
Joy, Bill, 139

Kant, Immanuel, 33, 91, 93-97, 107, 108, 111, 112, 255
Kaufman, Scott B., 49
Kelly, Jason J., 9, 26, 118, 258
Kelly, Kevin, 207
Kelvin, William T. (Lord), 20
Kepler, Johannes, 215

Koutroufinis, Spyridon A., 8, 51, 71, 87, 223, 246, 248, 281
Kurzweil, Ray, 127, 129, 130, 132, 137, 139, 144, 156, 171, 271, 275

Lapoujade, David, 80
Lloyd Morgan, Conwy, 249

MacCracken, Sean, 9, 61
Marx, Karl, 4
McCarthy, John, 138
McDonough, William, 239
McKenna, Terence, 252
Monod, Jacques, 20, 31
More, Max, 9, 160, 161, 271
Muhammad, 196
Musk, Elon, 139, 140, 147, 151-153

Naess, Arne D., 280
Nāgārjuna, 9, 155, 178, 179, 183
Nagel, Thomas, 173
Newton, Isaac, 4, 33, 71, 76, 94, 101, 102, 215
Nietzsche, Friedrich, 49, 75, 78

Parfit, Derek, 9, 155, 160-169, 171-178, 180, 181
Peirce, Charles Sanders, 13
Pikarski, René, 7, 105, 106, 114
Plato, 4, 36
Plotinus, 197, 226, 237

Rees, Martin, 139
Rickert, Heinrich, 108, 111, 115
Rosenberg, Alexander, 20, 31

Sagan, Carl, 135
Sahlins, Marshall, 30
Śāntideva, 174
Saussure, Ferdinand de, 13
Scheler, Max Ferdinand, 36-38
Schelling, Friedrich Wilhelm Joseph, 197
Segall, Mathew T., 6, 7, 110, 111, 120, 201, 243, 258, 263, 271
Seneca, 55, 56

Siderits, Mark, 174, 175
Smith, Adam, 4
Socrates, 36, 196
Spencer, Herbert Jennings, 249
Sri Aurobindo, 90, 237
Stevenson, Ian, 277
Suddendorf, Thomas, 39, 41
Swimme, Brian, 210, 226, 232, 237

Tarnas, Richard, 210, 221, 237
Thurman, Robert, 159

Uexküll, Jakob von, 8, 45, 79, 91-98, 107-110
Uexküll, Thure von, 92

Vasubandhu, 176, 272
Voegelin, Eric, 20

Wallace, Alfred Russel, 65, 215
Weinberg, Steven, 30
White, Lynn Jr., 185
Whitehead, Alfred North, 6, 8-10, 19-34, 37, 41, 59, 61, 65, 73, 75, 81, 82, 87, 89, 94, 96, 97, 102, 109, 111, 112, 114, 115, 118, 124, 130, 154, 187, 199-203, 210, 220, 222, 237, 247, 250, 253, 256, 259, 260, 264, 282
Whitman, Walt, 118, 187, 189-192, 194-196, 237
Wilber, Ken, 223

Index of Concepts

abstract concepts, 3, 97, 98, 108, 109, 113
abstract intelligence, 71, 104, 118, 120
Aborigines of Australia, 117, 119
absolute idealism, 158, 272
actual entity/entities, 81, 82, 257
actual occasion(s), 81, 87, 94, 97, 115, 259, 260, 264
  as flashes of existence, 259
  as physical-mental unities, 259
  as processes of experience, 259
  as proto-mental events, 260
  as synthetic acts, 259
  self-creation of, 259
Amazon (company), 151
Americans, Native, 119
Americas, 224
analytic philosophy/philosophers, 79, 89, 109, 256
animal
  intelligence, 8, 98-99
  species, 3, 107, 126, 142
animal rationale, 45, 98
animal symbolicum, 8, 45, 91, 98, 107
anthropic principle, 216, 217
anthropocene, 126, 265, 271
anthropocentric, 10, 26, 30, 185, 186, 202, 208, 213, 216, 218, 237, 263
anthropogenesis, 1, 2, 9
anthropologist(s), 19, 30, 45, 131
anthropomorphic, 30
apes, 12, 40, 151
apperception, 91, 93-97
  process, 93, 97, 98
  pure, 93, 95, 96
Apple (company), 146, 148
apprehension, 22, 50, 56, 236
  and prehension, 56
archetypal, 28, 121, 207, 208, 219, 221, 235-237
arrow of time, 86

artificial intelligence/AI, 127, 134, 138, 139, 167, 203, 210-212, 229, 267, 268
and human intelligence, 138, 139
strong, 167, 177, 182, 275, 276
atheism, 215, 217
augmented reality, 144, 148, 149, 229, 270
Australia, 119, 224
automate(s), 35
automatic/automatically, 50, 79, 125, 126, 147, 258
automaticity, 16
automation, 127, 137, 150, 228, 229, 239
automatization, 61
autonomous, 40, 61, 95, 112, 126, 135, 136, 140, 158, 167, 223
autonomy, 40, 54, 61, 82, 98, 136, 137, 140, 198, 215, 238
of the self, 54

Bagavad Gita, 189
biology
evolutionary, 64, 75, 93, 183, 248
evolutionary developmental, 66
of subjects 8, 93
philosophy of, 249
reductionistic, 21, 22
biomimicry, 235, 282
bionic human(s), 133, 134, 266
bio-philosophy, 39, 40, 43
Biopolitics, 272
biosphere, 2, 5, 43, 61, 110-112, 117, 119, 211, 217, 224, 225, 235
biotech, 8, 127, 132, 275
bio-technocratic/bio-technocracy, 2, 90
biotechnology, 109, 118
Bolivia, 120
brain(s), 6, 11, 14-17, 21, 22, 46, 88, 94, 97, 99, 120, 127-135, 137, 138, 140, 144, 145, 147, 148, 160-167, 175, 180, 183, 191, 211, 219, 220, 229, 275
digital global, 127, 144
synthetic, 211
Buddhism 9, 155, 157-159, 173, 174, 176, 177, 179-183, 237, 272, 277
and transhumanism, 157-159
Theravāda, 174, 274, 277
Tibetan, 157
Buddhist(s), 9, 61, 155, 157-159, 173-182, 220, 272, 273, 275, 277, 278
epistemology, 9, 157, 272
philosophy, 158, 176, 177, 180, 272, 273
Tibetan, 9, 155, 158, 178
Transhumanist-Dialogue, 61

capitalism/capitalist, 60, 148, 229
Cartesian
ego, 160, 168, 179, 180
self, 173
view, 176
causa sui, 81
causality, 12, 23, 70, 81, 91, 166, 178, 179, 257
and process philosophy, 81
efficient, 23, 81, 83, 257
final, 23, 257
formal, 23
causation, 13, 82, 166, 178, 268
efficient, 82
final, 82
centropic/centropy, 23, 31
Christian cosmology, 213
church, 214, 215
civilization, 2, 5, 21, 30, 31, 34, 117, 154, 186, 189, 191, 205, 269, 279, 280, 283
ecological, 2, 21
global, 213
Western, 5, 117, 213
closeness, 38, 43, 44, 47-50, 57, 59, 61, 95, 105, 111, 114, 115, 247
to life, 38, 43, 47, 48, 50, 59, 61

computer, 117, 127, 129-132, 134, 135, 137-140, 145, 147, 148, 151, 208, 265, 266, 268, 275
concrescence, 65, 81, 83, 97, 257, 259
concreteness
  misplaced, 66, 109
conditio humana, 2, 37, 38, 60
consciousness
  cosmic, 9, 10, 26, 118, 119, 185, 187, 188, 191-203, 237, 245, 280-282
  Earth, 234
  industrial, 234
  self, 95, 192
  super, 189
  symbolic, 25, 28, 110, 121, 232
  three stages of, 195
  universal, 188
  (ur), 112
continuity/continuities
  and difference 85
  evolution as heterogeneous, 7, 63, 71, 253
  heterogeneous, 7, 63, 69, 72, 74, 77, 78, 88, 248, 253
  indivisible, 70, 74, 251
continuous heterogeneity, 73, 85
continuum/continua
  concrete, 104
  heterogeneous, 71, 73, 81, 104, 113
  of numbers, 71
cosmic
  community of beings, 241
  consciousness, 9, 10, 26, 118, 119, 185, 187, 188, 191-203, 237, 245, 280-282
  evolution, 6, 20, 197, 209
  intelligence, 108, 282
cosmic subject
  universe as a great, 119, 218, 235
cosmocentric, 213, 218
cosmogenesis, 22, 24, 32, 119, 218, 219, 223, 224, 226, 242
cosmohumanism, 10, 206, 209, 210, 212, 213, 222, 227, 236

  and transhumanism, 210, 212
  as democratic phenomenology, 222
cosmohumanist, 10, 212, 213, 218, 221-224, 229, 230, 236, 242
  phenomenology 236
Cosmohumanists, 208-212, 218, 219, 228, 236, 239, 240
cosmology, 10, 22, 31, 33, 34, 206, 213, 217, 218, 221-223, 226, 227, 264
  Christian, 213
  of technology, 227
cosmomimicry, 234, 235
cosmos, 23, 26, 28-31, 34, 121, 187, 192, 195, 196, 198, 200, 202, 206, 212, 213, 216-218, 221, 226, 232, 234, 236, 238, 241, 245
  as immaterial and spiritual, 196
  psyche of the, 234, 238
creationism, 88
creativity, 3, 7, 30, 39, 47, 49, 74-79, 82, 83, 88, 96, 103, 127, 145, 148, 172, 201, 202, 209, 230, 231, 255, 257, 264, 276
culture, 14, 27, 29, 30, 33, 40, 100, 107, 117, 119, 124-126, 131, 148, 154, 207-209, 212, 215-217, 220, 221, 227-231, 233, 234, 237, 240-242
cyborg(s), 127, 133, 134, 140, 211, 266

DARPA (agency), 212, 228
Darwinism, 76, 91, 251
death, 19, 26, 34, 72, 80, 86, 87, 129, 131, 132, 137, 165, 166, 170, 183, 190, 191, 196, 210, 214, 274
deep
  ecology, 186, 199, 280
  psyche, 218
deity, 215, 226
democratic, 156, 189, 222, 226, 240, 275
designer babies, 127, 134, 140, 141, 143

determinism, 26, 86, 168
difference
  and continuity 185
digital, 110, 117, 118, 127, 130, 144, 145, 149, 210, 211, 230
  global brain 127, 144
  revolution 130, 145
divine, 34, 88, 119, 120, 188, 202, 213, 264
  Eros 202
DNA, 20, 88, 141-143
dream(s), 98, 219, 220
duration, 69-71, 74, 81-83, 85-87, 101, 103, 104, 112, 113, 119, 250, 255, 257, 264
  and life, 74, 83, 112, 250
  as heterogeneity, 71, 104, 119
  as heterogeneous continuity, 69
  as self-determination of essence, 81
durée, 42, 70, 246, 248
dynamical system(s), 88, 89

Earth, 1, 2, 5, 8-10, 12, 14, 23-26, 30, 50, 59, 98, 110, 119, 120, 125, 126, 145, 147, 148, 150-154, 187, 191, 192, 205, 206, 209, 210, 213, 214, 216, 221-223, 225, 227, 229-241, 271, 282
  as a living organism, 231
  as a subjective being, 231
  consciousness, 234
  Mother, 120
  Umwelt of the, 110
ecocentric, 186
ecological, 5, 9, 10, 14, 21, 25, 26, 30, 47, 59-61, 73, 110, 112, 114, 118, 185-188, 192, 198-203, 205-213, 223, 225, 228-230, 233-235, 240, 263
  civilization, 21
  crisis/crises, 5, 9, 10, 114, 118, 185, 198, 205, 240
  injustice, 192
  intuition, 118
ecology, 5, 9, 12, 110, 111, 120, 186,
187, 197-199, 202, 208, 223, 280
  deep, 186 199, 280
  spiritual, 197, 198, 202, 208, 280
economics, 109, 112, 114, 238
  neoclassical, 112
  neoliberal, 109
  process philosophical, 112
economy, 8, 30, 60, 111, 112, 114, 118, 146, 229, 239
  political, 30
eco-phenomenological, 227, 238
ecstasy, 189
Ecuador, 120, 264
efficient causality/causation/cause, 23, 81-83, 257
ego, 53, 69, 97, 104, 145, 160, 168, 179, 180, 216, 233, 237
  Cartesian, 168, 179, 180
  white male, 216
élan vital, 42, 79
eliminative materialism, 169
emanation, 264
embryo(s), 79, 80, 143, 255
emotions, 3, 17, 18, 49, 50, 98, 99, 131, 137
empathy, 24, 25, 39, 74, 79, 89, 209, 229, 230, 247
empiricism, 118, 200
Emptiness (Buddhism), 175, 176, 178, 179
engine 64, 77, 257
  universe as, 64
engineering, 68, 129, 134, 139, 141, 143, 156, 171, 207, 211, 228
  genetic, 134, 139, 141, 211, 228
Enlightenment, 36, 215
entheogenic substances, 237
entropy, 23
environment, 2, 5, 7, 8, 10, 12, 21, 27, 38, 44, 45, 47, 56-58, 91, 110-112, 121, 126, 128, 142, 146-148, 150-152, 157, 200, 211, 238, 259, 273
  sacred, 8, 110, 112, 121

epigenetics, 141, 142
epiphenomenon/epiphenomena, 23, 68, 96, 201
erotic, 25, 30, 100, 189-191, 196
 desire, 190
eschatology, 170
esoteric traditions, 220
essence, 7, 33, 38, 47, 59, 67, 71-73, 78, 81, 82, 85-87, 89, 95, 102, 104, 105, 109, 114-116, 119, 129, 167, 178, 179, 199, 208, 253, 254, 256
 determination of, 81
 interpenetration of, 82, 104
 intrinsic, 178
 transformation of, 72, 78, 81, 253
eternal, 32, 34, 196, 219
eternalism, 176, 180
eternity, 4, 255
ethical, 3-5, 8-10, 24, 26, 59, 61, 106-108, 110-112, 114, 118, 127, 128, 131, 135-137, 139-143, 159, 161, 174, 181, 182, 187, 192, 198, 202, 203
 imperative, 107, 110, 114
 Umwelt, 108, 111, 112, 118
 values, 3
ethics, 39, 58, 111, 131, 156, 157, 161, 163, 180-182, 191, 196, 198, 201, 202, 217
ethopoetic, 57, 59-61
ethos, 57, 58
eugenics, 143
Euro-American, 216
evolution
 as heterogeneous continuity, 7, 63, 71, 74, 253
 concept of, 63
 human, 1-8, 14, 17, 19, 20, 24, 29, 38, 99, 100, 105, 116, 118, 123-130, 133, 140, 144, 145, 150, 153, 181, 182, 190, 194, 206, 232
 of intelligence, 39, 59
 of species, 125, 249

 psychic, 219, 221, 231
evolutionism, 80
explanation, 1, 21, 23, 24, 31, 64, 67-69, 74-76, 83, 85, 87, 108, 109, 111, 161, 193-195, 200, 249, 257, 275
 as Procrustean bed, 67
external relations, 69, 81, 82
extinction of species, 126, 211, 224, 253
extraterrestrial life, 128, 153, 154

Facebook, 148, 212
faith, 3, 4, 120, 159, 170, 171
 naturalizing, 120
fallacy, 22, 26, 33, 68, 85, 109
 of bifurcation, 22, 26
 of misplaced concreteness, 66, 109
 of vacuous actuality, 33
final causality/causation, 23, 82, 257
finalism, 73, 80, 83, 118
 processual, 83
finalistic, 83
freedom, 3, 17, 35, 36, 38, 40, 41, 44, 46, 47, 51-54, 61, 71, 95-97, 111, 145, 162, 166, 167, 181, 191, 228, 238, 274
 and power, 38, 51-54
 morphological, 162, 166, 181, 274
 proto, 97
function circle/functional circle, 45, 92, 98
functionalism/functionalist, 9, 157, 169-171, 177, 179
futurism, 155, 156
 techno-, 156
futurists, 125, 137, 144, 154, 155, 157
 non-transhumanist, 157

game(s), 27, 51-54, 82, 89, 138, 148, 224
gender, 9, 157, 182, 188
genes, 109, 141-143, 228
genetic/genetically, 6, 11, 21, 22, 64,

84, 127, 133, 134, 139, 140-143, 208, 211, 218, 221, 223, 228
engineering, 134, 139, 141, 211, 228
mutants, 127, 133
genetically engineered human, 127, 140
genetics, 11, 14, 142
geocentric, 110, 214
German idealism, 191
God(s), 32-34, 66, 88, 100, 119, 130, 141, 143, 190-192, 196, 197, 200-202, 213-216, 226, 264, 282
  immanence of, 34
  incarnation of, 34
  Whitehead's, 34, 200, 202
Google, 129, 146, 148, 156, 212
gradualism, 248
Greek(s), 4, 89, 207, 214, 226
Greek Logos, 89

heliocentric, 214
Hermetic, 220
heterogeneity, 71-73, 82, 85, 90, 104, 106, 112, 119
  continuous, 73, 85
  duration as, 71, 104, 119
heterogeneous, 7, 52, 57, 63, 69-74, 77, 78, 81, 83, 85, 88, 104, 105, 113, 117, 248, 253, 264
  continuity/continuum, 63, 69, 77, 78, 81, 88, 104, 113, 248, 253
  duration as, 264
  evolution as, 7, 63, 71, 74, 253
  multiplicity, 71, 82, 83, 104, 248
high-tech, 5, 117, 118
Hindu(s), 159, 188, 220
Hindu philosophy, 188
Hinduism, 237
holism, 186, 199
holotropic, 237, 282
  states, 237
homoeroticism, 279
homogeneity, 71, 72, 77
  and essence, 72

homogeneous, 69-72, 78, 85, 90, 101-105, 117, 264
  continuum/continua, 71, 72, 104, 117
  discontinuity, 69, 72
  manifold, 72
homo sapiens, 24, 56, 145
human(s)
  bionic, 133, 134, 266
  condition, 7, 35-38, 43, 50, 54, 156, 183, 191, 197
  evolution, 1-8, 14, 17, 19, 20, 24, 29, 38, 99, 100, 105, 116, 118, 123-130, 133, 140, 144, 145, 150, 153, 181, 182, 190, 194, 206, 232
  freedom, 38, 47, 54, 95, 96
  genetically engineered, 127, 140
  intelligence, 38, 107, 112, 118, 119, 138, 139, 254
  nature, 2, 5, 12, 13, 17, 40, 74, 105, 117, 186, 187, 209
  species, 2, 4, 5, 22, 100, 105, 113, 129, 153, 206, 208, 231, 234, 242
  Umwelt, 51, 107
humanism, 9, 155
humanity, 2, 5-10, 23, 26, 36, 37, 61, 110, 119, 123, 124, 127, 128, 131, 134-136, 139, 140, 144, 145, 147, 150, 151, 153, 154, 182, 183, 190, 196, 205, 206, 210, 212, 214, 216, 218, 219, 227, 241, 258, 266
humankind, 4, 5, 23, 50
humanness, 6, 11-15, 18, 41, 42
  nature of, 13, 14

id-ego-superego (Freud), 145
idealism, 158, 191, 272
  absolute, 158
  German, 191
immanence, 34, 81, 197, 202
  of God, 34
  and transcendence, 197, 202
immaterial, 196
  cosmos as, 196
immortal, 131, 132, 168, 195

soul, 131, 132, 168, 195
immortality, 4, 5, 128-130, 132, 183, 191
imperative, 5, 54, 59, 68, 69, 106, 107, 110, 114, 198
  ethical, 107, 110
incarnation, 34, 211
  of God, 34
  of omnipotent AI entities, 211
India, 158
Indian, 158, 178
indigenous, 117, 119, 120, 125, 186, 216, 237, 272
  cultures, 119, 237
  peoples, 117, 216
industrial, 25, 127, 136, 206, 208, 217, 224, 225, 228, 233, 234, 241
  consciousness, 233, 234
  revolution, 224
  society, 208, 217, 224, 225
information, 3, 12, 16, 49, 127, 144, 145, 236
infotech, 8, 127, 129, 132
instinct, 41-44, 47-50, 53, 85, 113-117, 247, 262
  and intuition, 41, 42, 47-50, 53, 59, 85, 105, 113, 114, 116, 117
  as sympathy, 48, 49, 114
instinctive/instinctively, 43, 44, 47-49, 59, 61, 115, 165, 168
intellect, 7, 8, 10, 41-50, 53, 59, 60, 63, 65-72, 74-80, 82, 83, 85-90, 105-107, 117, 118, 196, 247, 249, 250, 252-254, 256, 257
intellectual, 4, 5, 7, 18, 42-45, 47, 50, 53, 59, 60, 65, 68, 69, 74, 76, 83, 89, 95, 118, 156, 157, 178, 179, 187, 195, 207, 214, 232, 247, 248, 250, 256, 273
intelligence, 8, 38, 39, 48, 59, 67, 69, 71-74, 77, 85-90, 98, 99, 102, 104-107, 112-114, 116-120, 127, 134, 135, 138-140, 143, 167, 203, 210-213, 217, 218, 229, 236, 237, 241, 254-256, 262, 266, 267, 282
  animal, 8, 98-99
  artificial, (AI) 127, 134, 138, 139, 167, 203, 210-212, 229, 267, 268
  cosmic, 108, 282
  human, 38, 107, 112, 118, 119, 138, 139, 254
  nature's, 106, 212
  of the universe, 106, 213, 236
  spatialized, 106
intelligent design, 88
intelligent species, 118
intentionality, 52, 165, 177
internal relations, 82, 114, 115, 259
Internet, 127, 128, 144-147, 208, 209, 229
intrinsic, 7, 8, 21, 22, 74, 81, 101, 106, 111, 112, 115, 116, 178, 186, 249
  essence, 178
  value(s), 8, 21, 22, 111, 112, 115, 116, 186
intuition, 2, 8-10, 27, 39, 41, 42, 47-51, 53, 59-61, 63, 69, 71, 74, 78-80, 85-90, 95, 96, 102, 105-107, 110, 112-114, 116-120, 221, 246-248, 250, 252, 276, 281
  age of, 8, 110
  and instinct, 41, 42, 47-50, 53, 59, 85, 105, 113, 114, 116, 117
  and the secrets of life, 116
  ecological, 118
intuitive/intuitively, 5, 7, 38, 48-51, 59-61, 73, 80, 89, 107, 116, 119, 121, 126, 186, 194, 196, 236
  technology, 61

Jaina, 159
justice, 21, 25, 112, 178, 197, 229, 238, 266

Kantian(s), 36, 41, 91, 95-97
Kantianism
  biological, 91
karma, 174, 176

language(s), 13-17, 39, 40, 78, 90, 94, 95, 97-100, 102, 109, 117, 120, 138, 156, 175, 222, 224, 231
libertarian, 156, 275
life extension, 129, 132, 156, 170, 171, 211
life world/lifeworld, 47, 50, 92, 108-110
linguisdic/linguistics, 12, 13, 15, 97, 99
logic, 16, 23, 24, 32, 66, 68, 80, 86, 87, 100
logos, 57, 58, 68, 74, 87, 89, 207
love, 34, 76, 118, 180, 190-192, 195, 196, 198, 202, 225, 241, 246, 255
LSD, 229

machine(s), 26, 32-34, 46, 88, 128, 130, 132, 134, 138, 140, 147, 156, 211, 216, 222, 223, 228, 241, 274
world, 34
Mahāyāna Buddhism, 174, 181, 182
mammals, 11, 15, 17, 24, 27, 151, 224
manifold(s), 45, 53, 72, 73, 91, 93-96, 98, 102
qualitative, 73, 102
mankind, 90
Mars, 8, 61, 110, 117, 147, 151, 152, 203
terraforming of, 110, 117, 151
materialism, 120, 169, 171
eliminative/eliminativist, 169, 171
scientistic, 120
materialist/materialistic, 9, 20, 31, 94, 100, 129, 159, 208, 209, 216, 219, 221, 222, 236
scientific, 216, 219, 222, 236
materialist(s), 30, 31-33, 132, 168, 171, 208, 220, 236
eliminativist, 171
scientific, 30, 31-33, 132, 168, 208, 220
Matrix, 211, 229
mechanical/mechanically, 24, 27, 35, 83, 102, 114, 133, 134, 215
mechanism(s), 16, 21, 22, 24, 33, 51-53, 64, 68, 73, 79, 82-84, 89
as metaphors, 68
mechanistic(ally), 46, 83, 84, 89, 101, 130, 199, 208, 213, 216, 217
memory, 3, 15, 16, 39, 103, 130, 229, 250, 255, 277
episodic, 3, 16, 39
procedural, 16, 39
symbolic, 103
mental, 3-6, 11-13, 15, 23, 39-41, 43, 50, 51, 59, 65, 71, 97, 103-106, 111, 115, 116, 169, 176, 220, 229, 234, 259, 260
pole, 97, 115
proto, 115, 260
mentality, 23, 115, 179
proto, 115
metaphor(s), 12, 13, 68
meta-physical, 115
metaphysical/metaphysically, 2, 10, 23, 24, 31, 33, 42, 48, 59, 115, 119, 170, 180, 198-202, 264, 272, 276
Microsoft, 139, 148
mind(s), 3, 6, 21, 23, 25-27, 39-41, 49, 55, 56, 67, 73-76, 80, 85-87, 97, 101, 102, 113, 117, 129, 130, 132, 145, 159, 162, 169, 181, 192-196, 199, 202, 220, 230, 237, 251, 256, 274
uploading 129, 130, 162, 274
misplaced concreteness, 66, 109
MIT, 129, 138
Monsanto (company), 228
morality, 2, 120
naturalizing, 120
morphological freedom, 162, 166, 181, 274
mortality, 4
multiplicity/multiplicities, 7, 63, 70-72, 74, 82, 83, 102, 104, 248, 251, 253

heterogeneous, 71, 83, 104, 248
qualitative, 82, 83, 248
quantitative, 82
mutants, 127, 133, 134
genetic, 127, 133
mystic(s), 217, 219, 220, 237, 258
mystical, 188-197, 200, 201
mysticism, 188, 193-195, 197, 280
myth(s)/mythic, 3, 5, 10, 25-28, 81, 98, 100, 110, 120, 121, 245, 263
mythmakers/mythmaking, 6, 10, 20, 33
mythopoeia, 7

nanotechnology/nanotech, 127, 139, 140, 143, 144, 149, 150, 211, 269, 275
NASA, 152, 270
Native Americans, 119
natural selection, 14, 21, 22, 64-66, 88, 120, 140, 141, 151
Naturalist(s), 208-210, 212, 227, 228, 236
naturalistic/naturalistically, 10, 22, 23, 32, 193, 199, 200
naturalizing faith and morality, 120
nature
  bifurcation of, 21, 29, 30, 33, 199, 201, 202
  human, 2, 5, 12, 13, 17, 40, 74, 105, 117, 186, 187, 209
  intelligence of, 106, 212
  intrinsic value of, 112, 186
  rights of, 120
neoclassical economics, 112
neo-Darwinism/neo-Darwinist, 76, 88, 91, 120, 249
neoliberal, 89, 109
  economics, 109
neo-liberalism, 82
Neoplatonists, 226
nervous system, 3, 6, 98, 106, 115, 220
neurophenomenology, 115
neurotheology, 120

new age, 186
Newtonian physics, 71, 94
nihilism, 31, 176, 179, 180
Nirvāna, 179
nonhuman citizens, 236, 240
non-reductionism, 175
non-reductionist/nonreductionist, 157, 168, 171, 180
noosphere, 145
nootropics, 229
novelty, 63, 64, 66, 69, 74, 82, 90, 250, 252

Omega Point, 164, 166, 274
ontological principle, 82
ontology, 24, 27, 34, 87, 89, 97, 114, 115, 118, 201
  panexperientialist, 27
  process, 89
  Whitehead's, 27, 34, 114, 115, 118, 201
organism(s), 6, 9, 21, 24, 26, 27, 30-34, 42-44, 47, 65, 72, 73, 79, 81, 86, 90-93, 96, 97, 102, 103, 111, 141, 142, 182, 185, 187, 199, 201-203, 207, 210, 211, 216-218, 222-226, 228, 231-233, 235, 236, 238, 249
  as a subject, 92, 222, 223
  cosmic, 218, 223, 235
  philosophy of, 9, 21, 81, 185, 187, 199, 201-203, 282
  universal, 224

Pachamama, 120
pagan/paganism, 186, 214
pan-en-theism, 197
panexperientialist ontology, 27
person(s) 9, 39, 51, 84, 120, 130, 133, 137, 143, 149, 157-169, 171-184, 194, 209, 256, 271, 274
personal, 5, 20, 32, 55, 57, 73, 119, 162-166, 168, 181, 194, 215
personal/person's identity, 163-165,

168
personhood, 9, 155, 160-162, 165-170,
    173, 174, 177, 180-182, 275
phenomenological, 10, 27, 29, 31,
    76, 206, 220, 227, 234, 236, 238,
    242, 282
  eco, 227, 238
phenomenology, 29, 108, 222, 236
  cosmohumanism as democratic,
    222
  cosmohumanist, 236
  theological, 29
philosophy
  of biology, 249
  of organism, 9, 21, 81, 185, 187,
    199, 201-203, 282
  process, 2, 81, 100, 112, 154, 173,
    201, 256
physicalism, 169
planet (Earth), 8, 10, 23, 25, 50, 59,
    61, 72, 73, 87, 100, 110, 118, 124,
    126, 128, 139, 145, 151, 153, 202,
    203, 205, 206, 209, 212, 217,
    218, 223-225, 227, 228, 231-235,
    237, 238, 240-242
  living technology of the, 232
play, 7, 24-30, 32, 34, 201
  divine, 34
playfulness, 7, 10, 25, 27, 98
poet of the world, 34
political economy, 30
post-human/posthuman(s), 125, 128,
    129, 139, 153, 210, 211
post-human species, 129, 153
posthumanist future, 212
postmodern, 60, 178
postmodernists, 208
power, 2, 4, 7, 10, 25, 31, 35, 37, 38,
    43, 47, 50-55, 58-60, 64, 75, 77,
    79, 93, 106, 110-112, 129, 138,
    180, 189, 203, 223, 226, 230,
    234, 235, 249, 266, 276
  and the Self, 53
  and freedom, 38, 51, 53, 54

as a processual concept, 52
Foucault's concept of, 37, 50
friendly face of, 51
relations, 2, 47, 54, 55
prehending
  entity, 81
  subject, 115
prehension, 115, 259
primates, 11, 24, 105, 142
process philosophy/philosophies, 2,
    81, 100, 112, 154, 173, 201, 256
  and causality, 81
process philosophical, 89, 112, 119,
    172
  economic and scientific theories,
    112
prophecies, 106, 111
prophetic future, 106, 107, 111
prophets, 106, 196
proto-
  freedom, 97
  justice, 25
  mental, 115, 260
  mentality, 115
  transhumanism/transhumanist,
    155, 182
psyche(s), 85, 211, 218, 228, 230, 234,
    236, 238
  deep, 218
  of the cosmos 234, 238
psychic/psychical, 11, 22, 83, 103,
    113, 207, 218-221, 223, 224, 231,
    235-237, 253
  evolution, 219, 221, 231
psychedelics, 208, 209, 229, 237
psychoactive drugs, 149
psycho-spiritual technologies, 235
punctuated equilibrium, 73, 248
purpose(s,) 22, 24, 25, 27, 35, 37, 39,
    40, 45, 46, 52-56, 61, 64, 65, 72,
    73, 80, 83, 86, 88, 111, 115, 118,
    140, 141, 148, 159, 162, 192, 197,
    198, 207, 213, 219, 226, 231, 232,
    241

purposeful, 32, 53
purposeless, 22, 33, 217
purposiveness, 23, 53

qualia, 96, 97
qualitative manifold, 73, 102
queer, 216

R-relatedness, 160-162
radical empiricism, 118, 200
radical finalism, 118
reason(s), 13, 16, 24, 41, 44, 47, 54, 59, 61, 80, 81, 88-90, 93-96, 153, 159, 167, 170, 228
reasoning, 2, 3, 14, 17, 90, 118, 158
rebirth, 26, 103
reductionism(s), 68, 84, 158, 161, 165, 168, 169, 171-175, 177, 180, 181
non-, 175
reductionist/reductionistic, 5, 21, 22, 118, 132, 157, 158, 160, 162, 165, 166, 168, 172-176, 178-180
non-, 157, 168, 171
reincarnation/reincarnated, 174, 182
Relation K, 174
Relation R, 160-163, 165-168, 172, 174, 178, 181, 183
relations
  external, 69, 81, 82
  internal, 82, 114, 115, 259
religion(s), 6, 7, 19-22, 24-30, 32, 33, 88, 98, 154, 155, 157, 159, 170, 185, 191, 193, 195-197, 208
religious, 7, 17, 19-23, 25-34, 106, 107, 119, 120, 157, 170, 183, 185-187, 189, 195, 196, 200, 201, 213-215
ritual(s), 18, 25-29, 34
robot(s), 127, 131, 132, 134-140, 211, 235, 266, 267
robotics, 127, 134-136, 138, 139, 151, 203, 228
romantic, 189, 194, 237
Romanticism, 191

sacred, 8, 26, 110, 112, 118, 119, 121, 190, 201, 217, 222, 234, 264
  and profane, 119, 190, 264
  environment, 8, 110, 112, 121
sacredness, 112, 119, 213, 236, 238
scientific materialist(s), 30-33, 168, 208, 209, 216, 219, 220, 222, 236
scientism, 6, 117
selection
  natural, 14, 21, 22, 64-66, 88, 120, 140, 141, 151
  sexual, 100, 260
  variation and, 64, 65, 74, 249
self/selves
  and its autonomy, 54
  as a human process, 54
  aware, 132, 135, 136
  care of the, 7, 38, 54, 55, 57, 59-61
  Cartesian, 173
  conscious, 48, 50, 116
  consciousness, 95, 192
  creation, 259
  determination, 81-83
  organizing/-organization, 23, 34, 111
  transcendent, 85, 116
  values, 111
semiosis, 92
semiotic, 6, 11-15, 92
sentient, 126, 132, 135-137, 181-183, 231, 267
sexual selection, 100, 260
shamanic, 219, 236, 237
  and transpersonal, 236
  mode of human reality, 237
shamans, 219
signs, 8, 17, 96-99
Silicon Valley, 61, 156
singularity, 127, 129, 145, 166, 275
social injustice, 192
socialism/socialist, 4, 148, 188, 275
Socratic, 207
soul 36, 103, 131, 132, 137, 142, 168,

        190, 191, 195, 202
    immortal, 131, 132, 168
space
    colonies, 110, 151, 152
    colonization of, 203
    stations, 151
space exploration, 124, 128, 135, 150, 154, 240
    and industrialization, 128, 150
    and settlement, 128, 150
Space X (company), 139, 147, 151, 152
spatialization, 70, 89, 100, 103, 105, 109, 111, 117, 119, 264
    of time, 103
spatialized intellect/intelligence, 72, 106
species 1-6, 10, 12, 14, 15, 18, 20-22, 24-26, 30-32, 40, 80, 86, 96, 98, 100, 105-107, 110, 113, 118, 120, 125, 126, 128, 129, 132, 135-137, 140-142, 151, 153, 154, 182, 205, 206, 208-211, 216, 223-225, 227-229, 231, 232, 234, 235, 242, 249, 253, 254, 263
    alien, 153
    animal, 3, 107, 126, 142
    evolution of, 125, 249
    extinction of, 126, 211, 224, 253
    human, 2, 4, 5, 22, 100, 105, 113, 129, 153, 206, 208, 231, 234, 242
    intelligent, 118
    posthuman, 129, 153
    symbolic, 6, 12, 18, 98
spirit(s), 65, 68, 130, 151, 152, 175, 188, 192, 226
spiritual, 2, 9, 19, 29, 30, 36, 55, 118, 129-132, 186-192, 196-198, 200, 202, 203, 214, 217, 219, 222, 230, 232, 235, 238, 239, 280
    ecology, 197, 198, 202, 208, 280
spirituality, 5-7, 19, 120, 186, 187, 189, 194, 197, 208, 209, 215
spontaneity, 48, 52, 59, 81, 88, 95, 96, 258

    transcendental, 96
STEM (science, technology, engineering, and mathematics), 156, 168, 271
subject(s), 7, 9, 32, 38, 51, 53-61, 92-98, 104, 105, 114-116, 119, 120, 157, 159, 163, 181, 187, 192, 193, 214, 216, 218, 222, 224, 226, 235, 259, 260
    animals as, 92-94, 96, 97
    biology of, 8, 93
    organism as a, 92, 222, 223
    universe as a great cosmic, 119, 235
subjectivity, 8, 53, 58, 59, 93-95, 97, 260
śūnyatā, 178 (*see also:* emptiness)
super
    consciousness, 189
    human, 133, 134, 211
    intelligent, 135, 136, 139
    intelligent AI, 139, 140
    natural, 26, 80, 200
    organisms, 238
    powers, 133
symbol(s), 6, 8, 12-14, 17, 25, 39, 45, 46, 51, 53, 56, 78, 83, 87, 96, 98, 99, 101, 105, 107, 109, 131, 230
    world of, 98
symbolic, 2, 3, 6, 8, 10-18, 25, 26, 28, 29, 39, 40, 45, 51, 88, 96, 98-103, 106-110, 113, 114, 117, 119-121, 206, 220, 222, 225, 232-234, 240, 242, 263
    consciousness, 25, 28, 110, 121, 232
    ecology, 12
    forms, 8, 26, 100, 106, 107, 113, 242
    future, 106
    memory, 103
    species, 6, 12, 18, 98
    system(s), 2, 3, 8, 26, 45, 99, 100, 107-109, 114, 117, 119, 120
    Umwelt(s), 8, 108
    universe, 45, 98

Index of Concepts

symbolicum
  animal, 8, 45, 91, 98, 107
symbolism(s), 25, 27, 99, 100, 102, 106, 108, 109
symbolization, 2, 6, 39, 99, 109
sympathy, 48-50, 60, 89, 114, 116
  instinct as, 48, 49, 114
synthetic brain, 211

techne, 58, 207
technocracy, 2, 5, 6, 89
technocratic, 5, 10, 74, 90, 111, 117, 118
Technologists, 208-211, 217, 227, 228, 236
Technology/technologies, 5, 8, 10, 17, 46, 61, 99, 100, 109-111, 114, 117-119, 123-128, 132, 134, 139, 140, 143, 144, 146-154, 156, 157, 160, 167, 170, 171, 183, 185, 203, 205-213, 216, 227-235, 239, 240, 271
  intuitive, 61
  of the planet Earth, 232
  psycho-spiritual, 235
Technoprogressive 156, 157, 272, 275
  Declaration 157, 272, 275
Technosophia, 10, 205, 213, 227, 234, 236
  as cosmology of technology, 227
teleology, 10, 34, 79, 84, 118, 119, 183, 213, 223, 226
  of the universe, 34
  participatory, 10, 119, 213, 223, 226
  transcendentalist, 183
teletransporter, 166, 168, 171, 172
terraforming of Mars, 110, 117, 151
Theravāda Buddhism, 174
thermodynamics, 23, 86
thought experiment, 163-165, 167, 168, 171
Tibetan Buddhism, 157
  and transhumanism, 157

Tibetan Buddhist, 9, 155, 158, 178
  philosophy, 158
trance states of consciousness, 219
transcendence, 32, 197, 202
  and immanence, 197, 202
  God's, 197
transcendental, 88, 94, 96
  evolutionism, 88
  spontaneity, 96
transcendentalism, 9, 170, 183
transcendentalist, 155, 170, 183, 189, 237
  teleology, 183
transgendered, 226
transhumanism, 2, 5, 9, 10, 38, 74, 90, 117, 118, 128, 129, 132, 156-159, 161, 169, 170, 172, 177, 179, 181-183, 210, 212, 213, 271
  and Buddhism, 157-159, 182
  and cosmohumanism, 210, 212
  classical, 161, 177
  hard, 9, 157, 172, 177, 179, 181
  proto-, 182
transhumanist(s), 118, 125, 128, 129, 132, 139, 143, 156, 161, 163, 169-171, 177, 178, 181-183, 211, 212
  and Buddhists, 178, 181
  hard, 177, 182
transhumanist, 6, 8-10, 61, 123, 124, 127-129, 132, 134, 155-158, 160-162, 169-172, 177, 179, 181-184, 210-213, 271, 275
  eschatology, 170
  hard, 9, 157, 160, 172, 177, 181, 183, 184
  philosophy, 155, 157, 158, 161, 162, 169, 177
  proto, 155
transhumanistic, 15, 16
transpersonal, 145, 197, 219, 220, 221, 230, 236, 237
  and shamanic, 219, 236
  dimensions of the universe, 236
  states of consciousness, 219

truth(s), 31, 57, 76, 89, 97, 119, 175, 178, 180, 181, 183, 201, 214, 215, 221, 223, 253, 273

Umgebung, 91, 97
Umwelt(s), 8, 51, 61, 91-93, 96, 107, 108, 110-112, 115, 118, 120
  of the Earth, 110
  symbolic, 8, 108
universal organism, 224
universe, 7, 10, 20, 22-24, 27, 30-34, 45, 63, 64, 69, 82, 83, 86, 98, 101, 106, 108, 118, 119, 123, 125, 126, 151, 164, 166, 191, 195, 196, 198, 201, 205, 208-210, 213-221, 223, 226, 227, 231, 232, 235-238, 240-242, 264, 282
  as a great cosmic subject, 119, 218, 235
  intelligence of the, 106, 213, 236
  symbolic, 45, 98
  teleology of the, 34
  transpersonal dimensions of the, 246
uploading of mind/consciousness/psyche, 127, 129, 130, 162, 211, 284
(ur)consciousness, 112

value(s), 1-3, 8, 16, 20-23, 27, 31-34, 42, 79, 108, 111, 112, 115, 116, 131, 157, 172, 176, 189, 201, 208, 213, 217, 220, 221
  ethical, 3
  intrinsic, 8, 21, 22, 111, 112, 115, 116, 186
  self, 111
variation, 64-67, 74, 81, 133, 212, 249, 251, 253
  and selection, 64, 65, 74, 249
virtual reality, 128, 144, 148, 149, 211, 229
vitalism/vitalistic, 66, 79, 88, 249

wisdom, 32, 137, 196, 212

www.ingramcontent.com/pod-product-compliance
Lightning Source LLC
Chambersburg PA
CBHW030317100526
44592CB00010B/464